Ultra-High Resolution NMR Methods

Upendra Singh

Table of Contents

Chapter 2

Parallel acquisition of slice-selective 1H-1H soft COSY spectra

Chapter 3

DQF J-RES NMR: Suppressing the singlet signals for improving the J-RES spectra from complex mixtures.

Chapter 4

Pure shift HMQC: Resolution and sensitivity enhancement by bilinear rotation decoupling in the indirect and direct dimensions

ABBREVIATIONS AND SYBMOLS

AQ	AcQuisition time
AP	Anti-Phase
BASHD	BAnd Selective HomoDecoupling
BIRD	BIlinear Rotation pulses and Delays
BIPs	Broadband Inversion PulseS
HOBS	HOmonuclear Band-Selective
BSE	Band-Selective Excitation
BSR-PSSE	Band-Selective Refocusing Pure Shift Spin-Echo
BURP	Band-selective Uniform Response Pure-phase; family of selective RF pulses: containing excitation (EBURP) and refocusing (REBURP) pulses
CPMG	Carr-Purcell-Meiboom-Gill
CT	Constant-Time
COSY	**CO**rrelation **S**pectroscop**Y**
CTP	Coherence Transfer Pathway
CPTCI	CryoProbe Triple-resonance Inverse with Carbon observe
CW	Continuous Wave
DMSO	DiMethyl SulfOxide
DSE	Double Spin-Echo
DOSY	Diffusion-Ordered SpectroscopY
DQF	Double-Quantum Filtered
DW	Dwell time

DS	Dummy Scan
E	Exclusive
EASY	Efficient Adiabatic SYmmetrized
FID	Free Induction Decay
FT	Fourier Transformation
HMBC	Heteronuclear Multiple Bond Correlation
HMQC	Heteronuclear Multiple Quantum Coherence
HSQC	Heteronuclear Single Quantum Coherence
IP	In-Phase
J	Scalar coupling between two spins
MQ	Multiple-Quantum
MRI	Magnetic Resonance Imaging
NMR	Nuclear Magnetic Resonance
NS	Number of Scan
NUS	Non Uniform Sampling
NOE	Nuclear Overhauser Effect
NOESY	Nuclear Overhauser Effect SpectroscopY
O1	Carrier Frequency in F_1 dimension
O2	Carrier Frequency in F_2 dimension
PASS	Parallel acquisition of slice-selective
PCA	Principal Component Analysis
PFG's	Pulse Field Gradients

PSYCHE	Pure Shift Yielded by CHirp Excitation
PSYCHEDELIC	Pure Shift Yielded by CHirp Excitation to DELiver Individual Couplings)
PE	Perfect-echo
PFG	Pulse Field Gradient
PK	Pell-Keeler
PS	Pure Shift
ppm	parts per million: unit describe the chemical shift
RASA	Rapid Absorption-mode and Strong-coupling-Artefactreduction
RGA	Receiver Gain
ROESY	Rotating frame Overhause Effect SpectroscopY
RF	Radio Frequency
SE	Spin-Echo
SITCOM	Stabilization by Interconversion within a Triad of COherences under Multiple refocusing
STOCSY	Statistical TOtal Correlation Spec- troscopY
SW	Spectral Window
SNR	**S**ignal to **N**oise **R**atio
TD	Time Domain
TOCSY	TOtal Correlation SpectroscopY
Z	Zero
ZQ	Zero-Quantum

ZS	Zangger-Sterk
t	Time
T	Temperature
t_1	Time axis of indirect dimension
t_2	Time axis of direct dimension
T_1	Spin-Lattice Relaxation time
T_2	Spin-Spin Relaxation time
ω_0	Larmor frequency in laboratory frame in rad s^{-1}
ω_1	Radio-Frequency
G_z	Magnitude of the gradient in the z-component of the field
γ	Gyromagnetic ratio

Chapter 1

1. INTRODUCTION

1.1 NMR Spectroscopy

Nuclear magnetic resonance (NMR) spectroscopy is a potent analytical tool to comprehend physical and chemical nature (mobility, dynamics and kinetics) of small to medium size molecules for an extensive range of samples under variety of conditions such as temperature, concentration, and pH. A wealth of information related to molecular properties and interactions can be furnished by using NMR which could consequently provide utility in molecular structural identification. However, low sensitivity along with low resolution is a concern in application of NMR. Within the past few decades, NMR sensitivity has improved significantly through advancement in instrumentation as well as methodological developments. Recent upgrade in NMR instrumentation such as cryogenically cooled probes[1] has led to increase sensitivity and three to four-times better signal /noise ratio in comparison to room temperature probes leading to faster acquisition times and improved sensitivity. Resolution of spectrum is additionally improved in a high magnetic field which disperses the chemical shifts over the broad frequency range (in Hz). Nevertheless, signal overlaps continue being a limiting factor for characterizing complicated spectra. Therefore, a steady development of new pulse sequences and enhancements of the existing ones are of vital importance in improving the overall performance of NMR spectroscopy.

In NMR spectroscopy, spin-spin coupling (also called scalar coupling or *J*-coupling) is an indirect interaction transmitted through chemical bonds between two spins[2,3]. Scalar coupling is one of the significant parameters in NMR spectroscopy for obtaining information on structure and dynamics of a molecule[2-4]. For example, dihedral angles from 2-bond proton-proton scalar coupling constants ($^3J_{HH}$) as per the Karplus relationship[2, 5, 6] could serve as an important information useful in structural and conformational studies for a molecule. Due to ease of utilization of proton spectra, ^{1}H-^{1}H coupling measurement is often used for molecular structural identification. Heteronuclear scalar couplings additionally give significant auxiliary data or fills in as a supplement to ^{1}H-^{1}H couplings and the NOE. Scalar couplings cause signal splitting (on account of weak coupling limit) and thereby coupling constants can be estimated from separations between significant peaks. In spite of the fact, the estimation of scalar coupling constants remains ambiguous. Overlapping of signals occur in proton spectra as a result of extensive signal splitting because of scalar coupling as well as narrow chemical shifts region (usually less than 12ppm). Therefore, the measurement of ^{1}H-^{1}H scalar coupling constants is frequently hampered by overlapping of signals and complex splitting pattern in conventional 1D ^{1}H NMR spectra. Complexities in the estimation of heteronuclear scalar coupling constants (J_{HX}) arise because of low sensitivity, complicated multiplet patterns, phase distortions and the coexistence of homonuclear and heteronuclear scalar couplings. Therefore, the accurate estimation of scalar coupling constants stays an indispensable job.

2D *J*-resolved experiment is an improved method to estimate scalar coupling constant with a greater ease[7]. Signal overlapping can be reduced by appropriate methods to disperse the chemical shift and *J*-coupling into two different dimensions thereby increasing accuracy of *J*-coupling constant measurements. However, peaks in regular 2D *J*-resolved experiments have to be displayed in magnitude mode due to phase-twist in line shape which causes a lowering of spectral resolution, reduction in accuracy of measurement of *J*-coupling constants and distortion of splitting pattern of peaks. A number of related techniques for computation of scalar coupling constants have evolved overtime[8] which prove their reliability on application in molecular structure analysis[9-11].

In past couple of years, much effort has been made to exclude homonuclear scalar J_{HH}-coupling constants from the ^{1}H NMR spectrum multiplets as an approach to eliminate signal overlap. The refocusing of scalar *J*-coupling impact in the conventional proton NMR spectrum produces a singlet at per chemical site. The spectral resolution increases due to a singlet signal which gives a definite position of chemical shift for a specific proton in a molecule. This refocusing of scalar *J*-coupling effect serves the purpose of broadband homonuclear decoupling, and this technique is known as "Pure-Shift" NMR Spectroscopy. There is a significant development of improved variants of Homonuclear broadband decoupling experiments in recent literature [12, 13].

One dimensional ^{1}H NMR is one of the most regularly used pectroscopic techniques to access qualitative and quantitative information from molecules. However, the information obtained from 1D ^{1}H NMR spectrum is limited, especially while being utilized for complex mixtures such as urine, saliva, plant extract, tissue sample, serum. 1D ^{1}H NMR method exclusively does not possess capability to identify all the metabolites in such complex biological samples because of overcrowding of spectral multiplet patterns.

Alternatively, different 2D NMR experiments can be useful in such circumstances where the signals are dispersed in two distinct dimensions (F_1 and F_2). Of all 2D NMR methods, COSY, and TOCSY have been widely used for identification of each metabolite among others using homonuclear as well as heteronuclear correlation. ^{1}H NMR spectra lacks resolution due to limited ^{1}H NMR spectral width which causes some overlapping signals to remain and develop complications in the analysis. There are a number of alternative methods which can be used to resolve this signal overlapping problem in 1D ^{1}H NMR spectrum. HSQC, HMQC and HMBC are heteronuclear experiments which elaborate chemical shifts of ^{1}H and ^{13}C nuclei in different dimensions (as F_2 and F_1 respectively) which reduces the signal overlapping enough to distinguish and identify peaks for a molecule in a complex mixture. Heteronuclear experiments offer far less sensitivity because natural abundance of ^{13}C is around 1.1% which is less compare to ^{1}H.

2. Correlation Spectroscopy (COSY)

A prevalent two-dimension NMR method is homonuclear correlation spectroscopy (COSY) experiment[14-15], which is utilized to recognize nuclear systems which are coupled to one another. It comprises of a solitary RF pulse (p1) trailed by the particular evolution time (t_1), further trailed by a second pulse (p2) and followed by acquisition time (t_2)[16]. The two-dimensional spectrum acquired by COSY NMR experiment demonstrates the frequencies for one type isotope, most ordinarily hydrogen (^{1}H) along the two axes. Two types of peaks appear in the 2D COSY NMR spectrum. Diagonal peaks with same frequency coordinates on each of the axes appear along the diagonal of the plot, while cross peaks have different chemical shift values for each frequency axes and are found off the diagonal scattered in the spectral region. Diagonal peaks correlate with the peaks in a 1D ^{1}H-NMR experiment, whereas cross peaks display the couplings between pairs of nuclei. Cross peaks occur due to phenomenon of magnetization transfer, and their occurrence reveals that two nuclei which have the two distinct chemical shifts that composing the cross-peak's coordinates are coupled. Each coupling results in two symmetrical cross peaks below and above the diagonal. Two cross-peaks are generated by the correlation between the two signals of the spectrum along each of two dimensions at these values. Fig. 1.1 shows the pulse sequence of 2D ^{1}H-^{1}H COSY NMR.

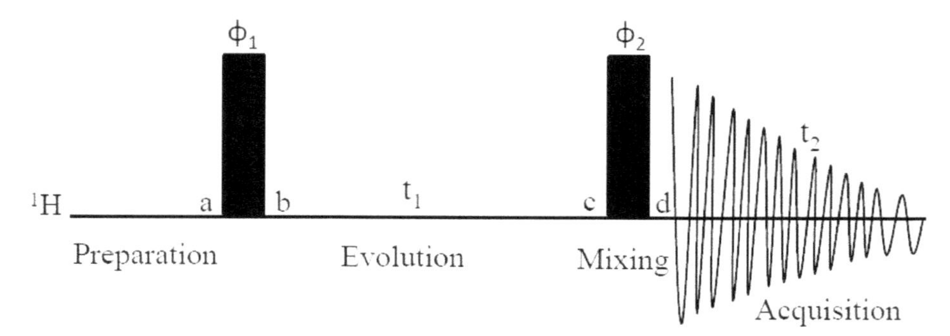

Figure 1.1. displays the pulse sequence of 2D ^1H-^1H COSY experiment. Both the filled rectangular bars are hard 90° pulses. First one is used for excitation of nuclei and second for mixing of their coherences.

2.1 Product Operators Involved in the Pulse Sequence

During the preparatory period of the pulse, magnetizations of nuclear spin systems get aligned z-axis. First hard 90° pulse about x-axis rotates all the magnetizations along y-axis as considered for binary spin system and thus the first magnetization commences. After first pulse, the magnetization evolves under both offset and scalar coupling. The last pulse mixes the coherences between coupling nuclei as a result.

$$I_{1z} \xrightarrow{(\pi/2)_x + t_1 + (\pi/2)_x} -I_{1z}\cos(\Omega_1 t_1)\cos(\pi J_{12} t_1) - 2I_{1x}I_{2y}\cos(\Omega_1 t_1)\sin(\pi J_{12} t_1)$$

$$+ I_{1x}\sin(\Omega_1 t_1)\cos(\pi J_{12} t_1) - 2I_{1z}I_{2y}\sin(\Omega_1 t_1)\sin(\pi J_{12} t_1)$$

$$\cos(\Omega_1 t_1)\cos(\pi J_{12} t_1) = \tfrac{1}{2}[\sin(\Omega_1 t_1 + \pi J_{12} t_1) + \sin(\Omega_1 t_1 - \pi J_{12} t_1)$$

$$\cos(\Omega_1 t_1)\sin(\pi J_{12} t_1) = \tfrac{1}{2}[\cos(\Omega_1 t_1 + \pi J_{12} t_1) + \cos(\Omega_1 t_1 - \pi J_{12} t_1)$$

The first and second terms out of the all four terms are zero and double quantum coherence respectively which can not be detected by spectrometer, but third and fourth terms are single quantum coherence which are detected. Third and fourth terms give diagonal peak and cross peak respectively which are phased by 90° simultaneously. Therefore, they cannot be observed in phase mode.

Fig. 1.2 (a) shows the molecular structure of L-methionine and Fig. 1.2 (b) shows its 2D ^1H-^1H COSY NMR spectrum in the magnitude mode with assigned peaks. The 2D COSY spectrum shows the correlation among signals which are separated by three bonds as signal 2 correlates with the signals 1and 3. These correlations result in the formation of rectangular by the cross peaks and diagonal peaks of coupling nuclei as shown in Fig 1.2b).

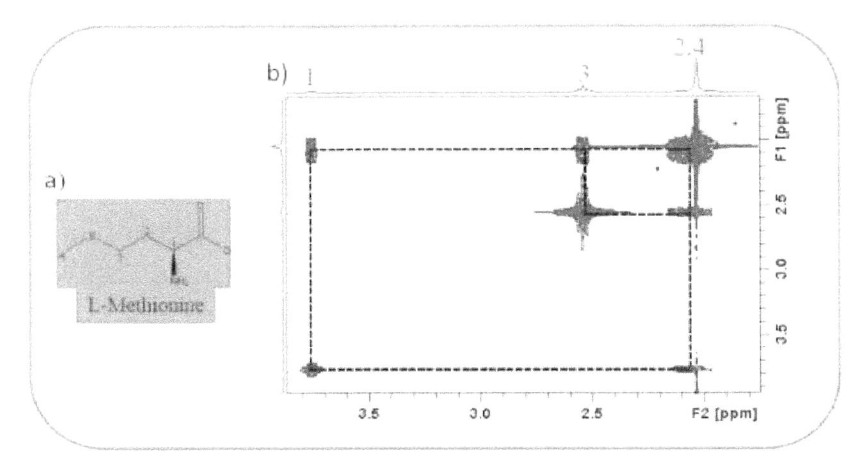

Figure 1.2. displays the molecular structure of L-methionine in Fig. 1.2a) and its 2D NMR ^{1}H-^{1}H COSY spectrum with assigned signals in Fig. 1.2b).

There is no formation of rectangle between signals 1 and 3 because they are separated by four bonds. Therefore, there is no bond correlation between each other. The diagonal peaks have large tails which can obscure the peaks closer to it causing reduction in resolution. The high intensity of singlet peaks also obscure nearby signal peaks. Therefore, this experiment needs to be improved for resolution.

2.2 Demerits of 2D COSY NMR

As there are many advantages to the 2D COSY NMR methods, several drawbacks of the current method employed also need to be considered in order to improve the efficacy of this experiment. Firstly, the anti-phase of the cross peaks leads to spectral lines cancelling out one another, and the in-phase structure of the diagonal peaks, which boosts the peak intensity giving a false ratio and hampering quantitation. Second, there is a significant difference in intensity between the cross peaks and the diagonal. This difference in intensity causes difficulty in identifying small cross peaks, particularly if they lie close to the diagonal. Another important concern is the broad lineshapes incorporated while processing the data for a COSY spectrum which reduce the spectral resolution further.

2.3 Application of 2D COSY NMR

Two-dimensional NMR correlation experiment as COSY finds utility in structural elucidation of gaseous analytes[17]. These gaseous analytes may be composed of hydrocarbons or fluorocarbons. ^{19}F-COSY NMR is very known and reliable experiment for analysis of fluorocarbon. Acquisition time for this technique is faster than other 2D-NMR experiments. It provides correlation between fluorine nuclei and helps in the interpretation of the surrounding chemical environment[17]. 2D ^{1}H-^{1}H COSY NMR is useful for investigation the intermolecular interactions and to get further information on the confirmation of the inclusion complexes [18]. 2D-COSY NMR has proven its utility in the quantification of biological matrix (2-hydroxypropyl-β-cyclodextrin in plasma) with valid statistical analyses like linearity, trueness, precision, limits of quantification and accuracy[19]. ^{1}H-^{1}H couplings are determined by fitting the cross-peak of multiplets in 2D regular phase-sensitive COSY spectra[20].

3. Soft ^{1}H-^{1}H COSY

This method extracts a cross-peak corresponding to a signal resonance frequency which couples with our selectively excited signal to reduce the crowding and provide the facility to measure coupling constant accurately between those two coupling signals. The soft ^{1}H-^{1}H COSY experiment consists of three semi-selective pulses, first for excitation, second for restoring and a final pulse for mixing. The latter two pulses allow a coherence transfer only between the limited spectral range $\Delta\Omega_1$ and $\Delta\Omega_2$. The pulse sequence of the soft ^{1}H-^{1}H COSY[21] is shown in figure 1.3.

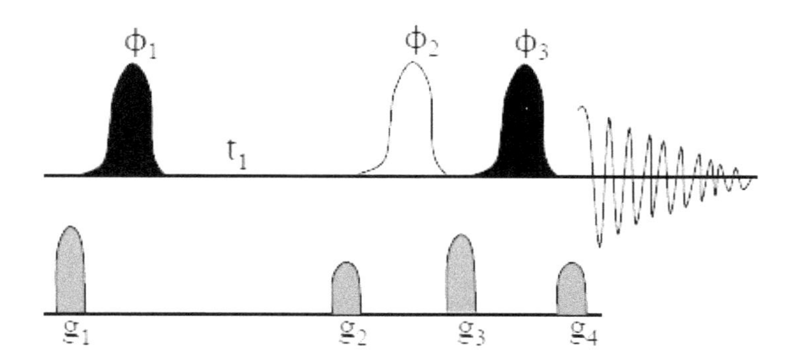

Figure 1.3. the pulse sequence of soft ^1H-^1H COSY shows selective 90° pulses depicted by the shaped hollow and filled bars. g_1 and g_3 are sine shaped gradient for crushing. g_2 and g_4 are sine shaped gradients for coherence selection gradient. Pulse sequence of soft ^1H-^1H COSY with first two selective 90° pulses excites the I_1 multiplet and the last mixing 90° pulse excites the I_2 multiplet of a three-spin system (I_1, I_2, I_3).

A resemblance to the basic sequence of the heteronuclear ^1H-X COSY experiment[22] is evident. Theoretically, the last two selective pulses for the coherence transfer should be applied simultaneously, but practically this could not be achieved due to software and hardware restrictions. The first two pulses have to be applied to spin 1, whereas for the last pulse the transmitter frequency has to be switched to the centre of the spin 2 multiplet representing the frequency of detection. The soft ^1H-^1H COSY experiment provides cross peaks shown in Fig. 1.4 like an E-COSY[23] multiplet pattern. The soft ^1H-^1H COSY spectra for (S)-4-methyloxetan-2-one (β-butyrolactone) is shown in Fig. 1.4. The proton 2 in β-butyrolactone exhibits homonuclear scalar coupling with the proton 1 and 3 as shown in the molecular structure (ignoring long range scalar coupling $^4J_{HH}$ to signal 4). The signal 2 is selectively excited by the first two selective pulses, so that, chemical shift and scalar couplings of signal 2 is encoded in the F_1 dimension. In the end, the mixing pulse can be applied on the desired signal 1 or 3 according to acquire cross-peak at a desired chemical shift of the peak signal. The mixing pulse is applied on peak 3 therefore method provides cross peak corresponding to the chemical shift peak 3 and its expansion as shown in Fig. 1.4A and 1.4B respectively.

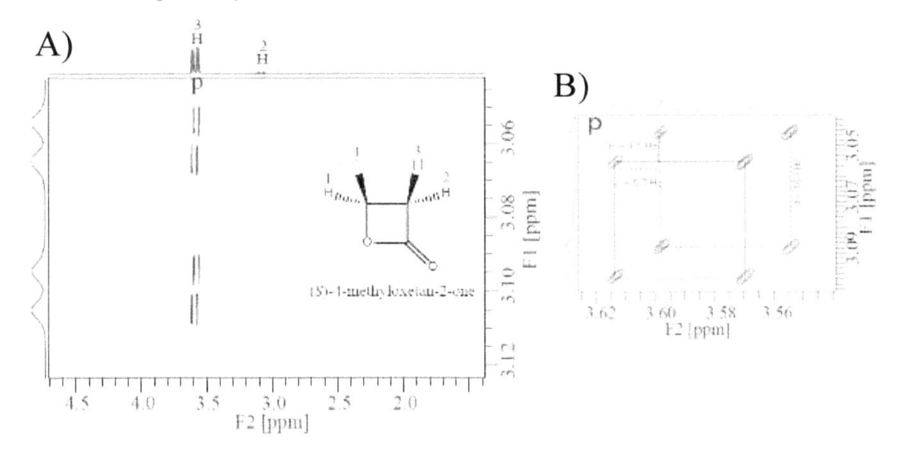

Figure 1.4. displays the 2D spectra of soft ^1H-^1H COSY method of (S)-4-methyloxetan-2-one recorded by the pulse sequence shown in Fig. 1.3. One cross-peak is acquired through soft ^1H-^1H cross peak which is similar E-COSY multiplet pattern.

The cross-peak of the soft COSY spectrum (Fig. 1.4B) contains two types of coupling. Active coupling a = $^{2}J_{H2H3}$ (=16.3 Hz) along both dimensions shown as a regular square in Fig. 1.4 which is a result of coupling between two selectively excited signals 2 and 3. Passive coupling c = $^{3}J_{H1H3}$ (=5.7 Hz) leads to displacement in the square along F_2 and thus the coupling constant value can be measured along F_2. Similarly, the other passive coupling b = $^{3}J_{H1H2}$ (= 4.1 Hz) leads to displacement of the square boxes along F_1 and thus, value of passive coupling constant can be measured along F_1. The active coupling has an anti-phase line-shape, while the line shape due to passive coupling occurs in-phase. Thus, the soft ^{1}H-^{1}H COSY provides a simpler 2D spectral pattern for extraction of the scalar coupling.

3.1 Drawback of soft-COSY Method

It can measure only single active coupling constant between a selected pair in a single experiment. As a result, a huge set of experiments have to be run for estimating number of coupling constant among coupling protons. An improved version of this experiment is needed to provide all coupling constants of a selected proton in a single experiment which is discussed in the next section.

4. Zangger and Sterk (ZS) Method

Klaus Zangger and Hein Sterk developed a technique for homonuclear decoupling by slice-selection in liquid NMR spectroscopy which allows the spatial separation of various nuclear spins with different resonance frequencies in equally sliced sections of sample contained in NMR, known as slice-selection, as shown in Figure 1.5 (a). Slice selective selection excitation is obtained by using frequency selective pulse as well as weak pulse field gradient (PFG) simultaneously which causes the position-dependent shift across the sample of NMR tube. These combined selective pulse and weak gradient is called ZS pulse[24-25]. The spatially selective pulse excites all the signals of different slices at the same time[24-27] . Similarly one can employ spatially selective refocusing pulses by which the possibility of scalar homonuclear coupling does not occur among the signals shown in Fig 1.5 (b). Homonuclear decoupling enhances resolution in signals. Spatially-specific excitation has been utilized regularly for in vivo MRI applications[28], and is being constantly improved to find its use in liquid NMR applications[29-35]. In an isolated slice, the excited signal with refocused *J*-evolution can be obtained by applying an additional slice-selective 180° pulse as well as a non-selective 180° pulse shown in Figure 1.5(c).

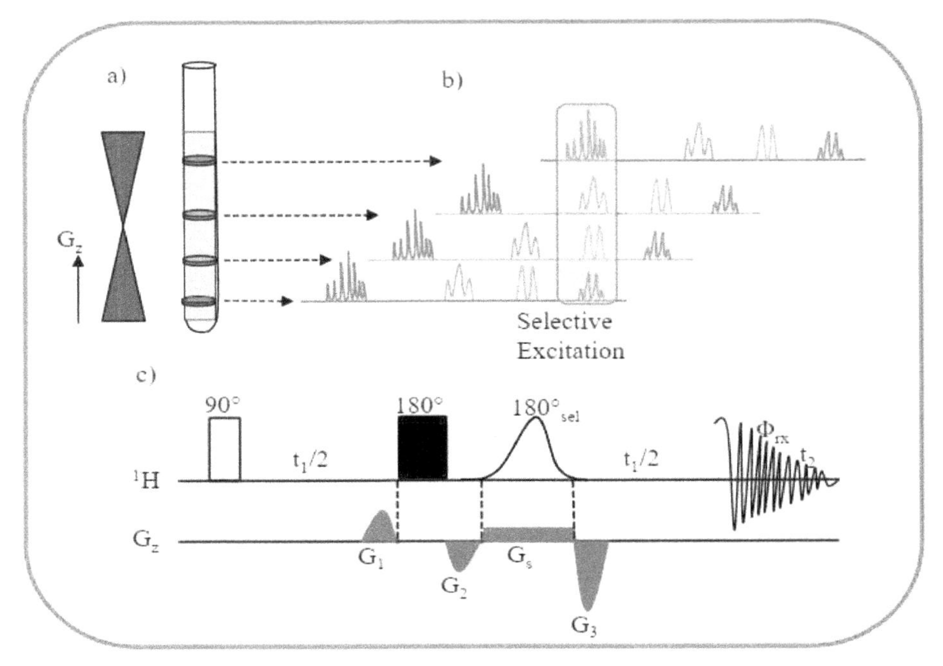

Figure 1.5. (a) displays the principle of spatial position-dependent shift across the NMR tube due to applied weak gradient G (Gauss/cm) along z-axis, b) schematic diagram of slice selective excitation of signals in different slices and c) pulse sequence of Zangger-Sterk pure shift method.

The status of the signal which is on resonance within this slice remains same, however, other signals belonging to off resonance are inverted. Scalar coupling evolution before the refocusing pulses is subsequently refocused. This decoupling technique might be handily executed in the indirect dimension of multidimensional experiments[24]. In order to fulfil this objective, the decoupling pulses must be situated in the middle of the evolution period. The sensitivity of slice selective decoupling methods is considerably reduced compared to the regular proton spectra because each signal is excited in a thin section of the active volume of sample tube. The sensitivity of this method also depends on the bandwidth of the slice-selective excitation pulse.

A fraction of the reduced sensitivity can be recovered by decoupling. The excitation bandwidth of slice-selective pulse needs to be optimized according to vicinity of the scalar coupled protons in the spectrum. Highly selective pulses need to be applied if chemical shifts of two protons are very close to each other. The sensitivity of the spectrum decreases with lowering of excitation bandwidth. Therefore, sensitivity becomes limiting factor for this method and thus it is less suitable for decoupling in case of strong coupling. The central induced (shifted) frequencies (in Hertz) of slices depend on length of active volume sample s (in centimetre), strength of applied weak gradient G (Gauss/cm) and gyromagnetic ration γ as $\Delta\upsilon = \gamma*G*s$. The signal to noise ratio (S/N) is directly proportional to the bandwidth of selective pulse as S/N is $\Delta\omega/\gamma*G*s$, where $\Delta\omega$ is the bandwidth of selective pulse. Slice-selective decoupling method has been derived in the indirect dimension of two-dimensional homonuclear NOESY[36] and TOCSY [37, 38, 24] experiments and homonuclear ^{13}C-^{13}C decoupling by slice-selective ω_1-decoupling has also been achieved for uniformly ^{13}C-labeled compounds[39-40]. The dispersion components in *J*-resolved spectra can also be removed by applying slice-selecting refocusing[41].

4.1 Application of ZS element in the gradient-encoded homonuclear selective refocusing (*G-SERF*)

The sequence of this method aims to provide a simplified method to assign and measure all the homonuclear couplings between a chosen proton and its adjacent coupling protons in the spin system by using spatial encoding of frequency in slices to *J*-edited spectroscopy[42] acquired in a single 2D experiment. The G-SERF method originates from the SERF method which measures the only single homonuclear coupling constant between a pair of couples signals in a single experiment[43]. The G-SERF method is based on spatial frequency encoding element for extracting desired coupling constants. The pulse sequence of G-SERF method is shown in Fig. 1.6 which consists of ZS element. G-SERF has advantages over SERF as it is easier to extract all scalar coupling constants involving a desired proton within the coupling network at once, notably enhancing the efficiency of the SERF experiment. The selective 180° pulse with the weak filed gradient simultaneously inverts different protons in different slices of sample due to weak field gradient which splits the spins present in the slices. Only homonuclear scalar couplings of selected protons evolve during evolution time t_1, while all chemical shifts and rest of the couplings are refocused. Therefore, signals of protons which couple with the selected proton are split into doublets along the F_1 dimension. In addition, the z-filter is applied before acquisition to get a spectrum in absorption-mode and improve the resolution as well.

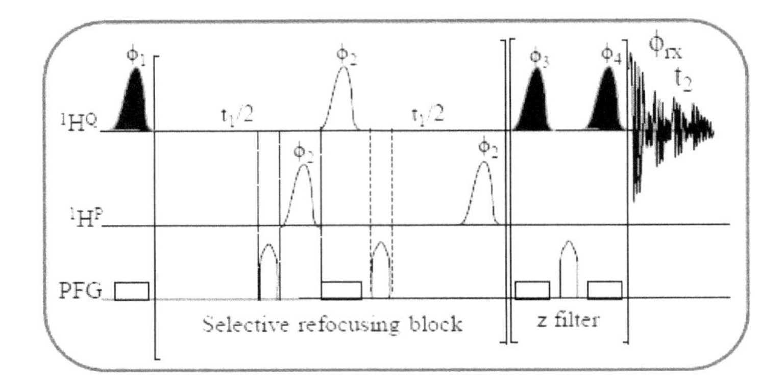

Figure 1.6. displays the pulse sequence of G-SERF method. Black and white half sine shapes on the proton channels $^1H^Q$ and $^1H^P$ are selective 90° and 180° pulses, respectively. Non-encoded selective 180° pulses are set to selected proton (P).

The utilization of the spatial frequency encoding element leads to a significant loss in sensitivity of signals in G-SERF spectrum because signals are produced from different slices across the whole active sample, similar to pure shift obtained from ZS experiments. Therefore,

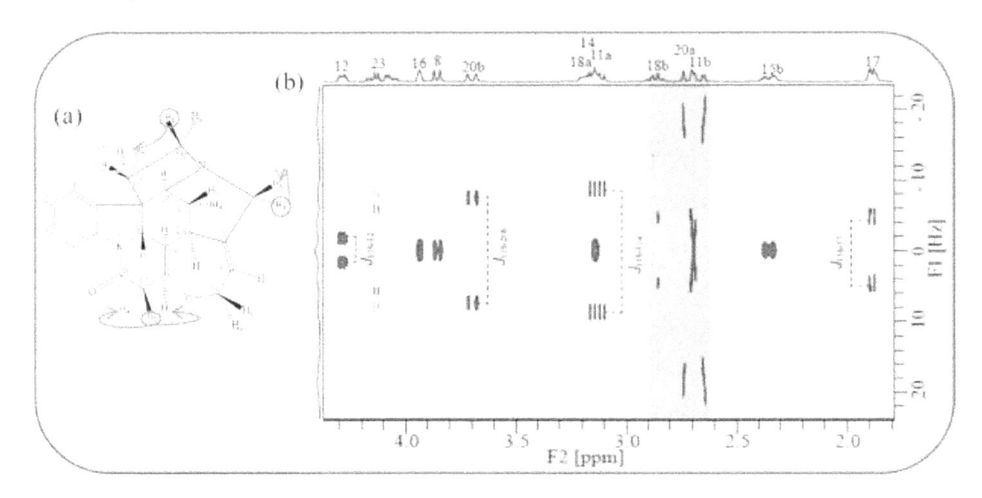

Figure 1.7. (a) represents the molecular structure of strychnine, (b) represents the 2D ^1H G-SERF spectrum of strychnine in $CDCl_3$ solvent. Non-encoded semi-selective 180° pulses are set to excite the resonance frequencies of the proton spins H^{11b}, H^{18b}, H^{20a}.

magnitude of signal intensity is directly proportional to thickness of the slice, value of which is related to the bandwidth of the selective 180° pulse and the frequency encoding gradient. The signal to noise ratio is directly related to the bandwidth of selective 180° pulse, therefore it is set to a maximum optimized value. For a sample with crowded signals, the bandwidth of the selective 180° pulse needs to be narrow enough to avoid interference among slices and invert different protons separately, hence resulting in the loss of sensitivity.

5. Two-Dimensional *J*-Resolved Method of NMR Spectroscopy

J-resolved experiment[44-49] of NMR spectroscopy is the first method to achieve decoupled proton spectra developed by Ernst and co-workers in 1976. This experiment works based on the spin-echo[50] and distributes the chemical shift and scalar coupling information in two distinct dimensions. Thus, analysis of any one of these parameters could be accomplished without interference of the second. Scalar couplings are represented in F_1 and chemical shift in F_2. The pulse sequence of this experiment has been shown in Fig. 1.8. This pulse sequence (figure 1.8a) consists of 180° hard pulse at the centre of evolution period t_1, to produce a spin-echo. Spin echo is adjusted by homonuclear *J*-coupling[51] which results in the distribution of signals along the indirect (F_1) dimension of the observed 2D *J*-Resolved spectrum. It causes the separation of *J*-couplings from chemical shift, amplifying the spectral resolution for individual signals in a spectrum with overlapped multiplets.

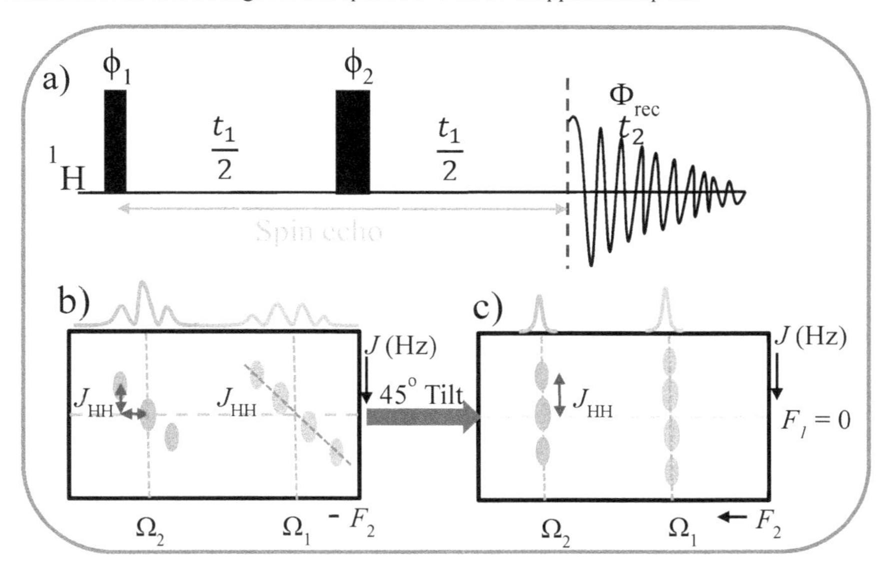

Figure 1.8. a) displays the pulse sequence of *J*-resolved experiment. Narrow filled black and wide filled black bars represented the hard 90° and 180° pulse respectively. Schematic 2D *J*-Resolved spectrum with multiplet signals which are tilted by 45° from the horizontal axis $F_1 = 0$ (Hz) and aligned along F_1 axis shown in Fig. 1b-c respectively.

It also suppresses the B_0 inhomogeneity effect which results a further gain in spectral resolution along F_1 dimension. This spin echo modulation includes dispersion-mode component in the 2D *J*-Res spectrum which causes phase twist lineshape. These undesirable phase twist lineshapes are composed of a mixture of absorptive and dispersive mode lineshape. The spectrum is introduced as a 2D diagram where signals are centred at their chemical shift in F_2 dimension following a tilt of 45° to the F_1 dimension Figure (1.8b). The signals without a tilt by 45° around their mid-point make the multiplets align the F_1 axis and allows observing singlets in the F_2 projection Figure 1.8(c).

Figure 1.9 (a) shows the 2D *J*-resolved NMR spectrum of strychnine molecule sample. The signals in spectrum are tilted by 45° from the horizontal axis $F_1 = 0$ (Hz). Therefore, the spectrum shows the multiplets along F_2 dimension.

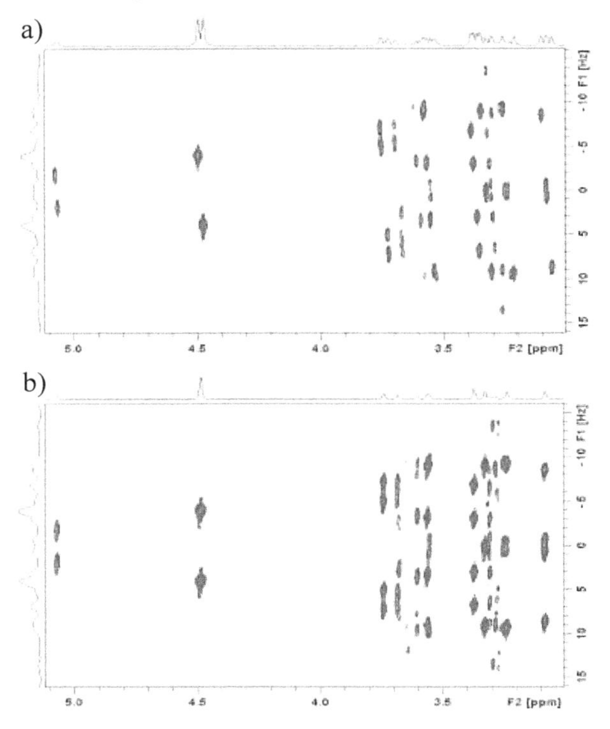

Figure. 1.9. Pulse sequence of *J*-resolved was used in recording of spectrum of strychnine molecule in $CDCl_3$ solvent. (a) *J*-resolved spectrum with tilt of 45° from the axis of F_1=0. (b) *J*-resolved spectrum without tilt of 45° obtained after applying tilt.

In the Figure 1.9 (b), spectrum shows singlets of signals along F_2 after applying tilt and signals are aligned straight along F_1 axis. Therefore, spectrum represents 1D pure shift along F_2 dimension which leads to enhancement of resolution in signals but spectrum of *J*-resolved experiment suffers from phase-twist lineshapes in signals which leads to broad singlet signals along F_2. At present, few reliable techniques are available which remove dispersive part from signals and overcomes this problem.

5.1 The methods for removing phase-twist in lineshape of *J*-resolved spectra

5.1.1 The method for absorption-mode of *J*-resolved spectra

The method for acquiring absorption-mode *J*-spectra depends on the notable idea that a P-type and an N-type spectrum both when added properly provide an absorption-mode spectrum[52-53]. Both the P-type and N-type spectra have phase-twist lineshapes, but if the ω_1 axis of one of the spectra is changed, and both the spectra added together, it causes the dispersive parts of

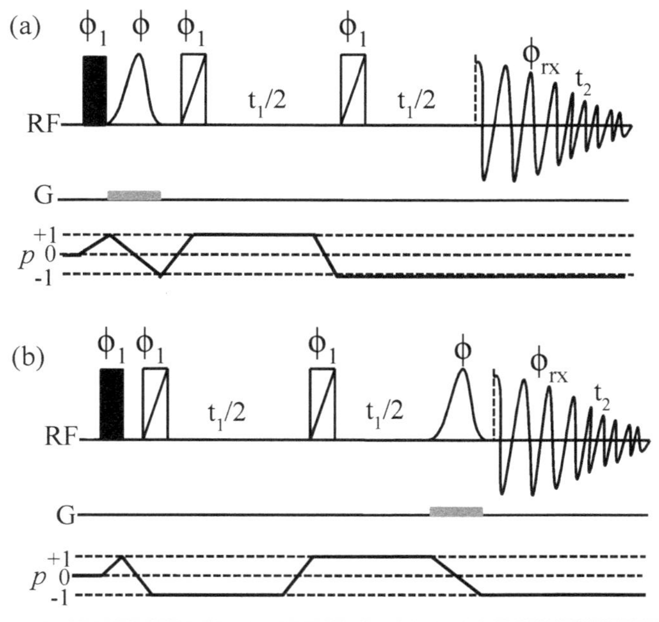

Figure 1.10. displays pulse sequences of conventional *J*-Resolved in (a) and anti *J*-resolved in (b) along with coherence transfer pathways. Filled black rectangles and unfilled rectangles with diagonal strokes represent the hard 90° pulses and BIPs respectively, while half sine shapes represent the selective 180° pulses.

the phase-twist lineshapes to cancel out. The pulse sequence of the conventional *J*-resolved experiment as well as the coherence transfer pathway (CTP) has been shown in Fig 1.10(a). Indeed, the P-type or N-type spectra cannot be distinguished because the coherence order changes in the middle of t_1. The pulse sequence of complementary *J*-res spectroscopy along with involved in its CTP has been shown in Fig. 1.10(b) which has a phase-twist in opposite sense. In this sequence, the coherence order is -1 during the first half of t_1, and +1 during the second half, reverse to the CTP in Fig. 1.10(a). Simply, we can reverse between the P-type and N-type CTP by changing the gradient pulses or phase cycling, but there is no way to reverse between CTPs for the *J*-spectra and anti *J*-spectra. *J*-res has no mixing pulse at the end of t_1, therefore a pulse has to be put in the end to achieve desired coherence order (-1) for recording spectra in case of Fig. 1.10 (b). In fact, a pulse can also flip the spin states of all passive spins. As a result, it yields signal in a manner that the dispersive parts of the lineshapes do not cancel when the *J* and anti *J*-spectra are added. The desired result could be achieved only by applying a selective 180° pulse to single spin at a resonance, such that the spin states of the entire passive spins are remain unchanged. The selective 180° pulse is replaced by Zangger-Sterk[54] pulse sequence element at the end of t_1, as shown in Fig 1.10(b). This element contains a selective 180° pulse associated with a weak gradient simultaneously. The strength of gradient is chosen such as the spread of frequencies across the sample matches the spectral width.

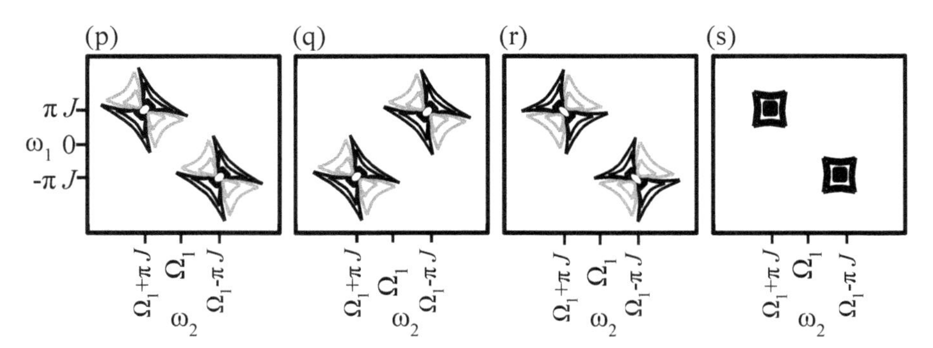

Figure. 1.11. One of the two-spin system is shown in a simulated spectrum. The spectrum of conventional *J*-Res is shown in (p), and spectrum of anti *J*-Res, in which the doublet tilt in the opposite direction is shown in (q).

The peaks have the specific phase-twist lineshape. The ω_1-axis of the spectrum in (q) is reversed to achieve (r), which is then added to (p) to give the absorption-mode spectrum in (s). Negative contours are grey, and positive contours are coloured black.

The absorption-mode spectrum of a normal two-spin system is produced by outlining with simulations, as shown in Fig. 1.11, The conventional *J*-spectrum is shown in fig. 1.11(p), the anti *J*-spectrum is shown in (q), and (r) is spectrum (q) with the ω_1-axis reversed. The absorption-mode spectrum results by addition of (p) and (r) is shown in 1.11(s).

5.1.2 Pseudo echo processing of *J*-Res spectra

A general method of eliminating dispersion-mode components from a free-induction decay is known as pseudo-echo processing[55]. This scheme derives from the observation that when appropriately phased, the two halves of a spin echo consist of the absorption-mode components in the same phase but the dispersion components in antiphase. The antisymmetric components produce the dispersive part after applying Fourier transformation and eliminated from a time-domain function which decays in a symmetrical form on both sides of the midpoint shown in Fig. 1.12(c). Thus, a pseudo-echo can be generated by first multiplying a window function $\exp(+t/T_2^*)$ with the free induction decay (FID), thereafter multiplying by a Gaussian envelope to yield the symmetrical pseudo-echo. After 45° projection of the 2D *J*-resolved spectrum, remote peaks show up in absorption mode, however distortions stay wherever overlapping of peaks occur. The sensitivity of the resulting pure shift spectrum is considerably decreased since the first points of the FIDs are greatly reduced in intensity.

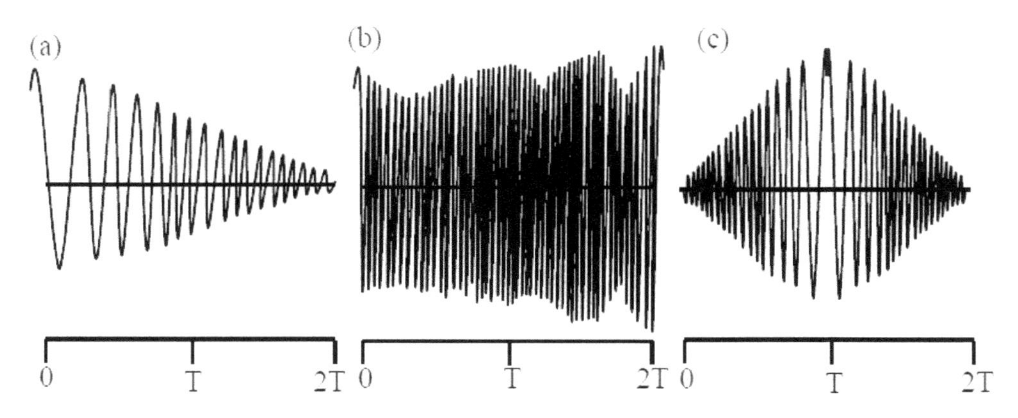

Figure 1.12. (a) displays free-induction decay of an experimental signal with an exponentially decaying envelope. (b) The same signal multiplied by $\exp(+t/T_2^*)$. (c) A "pseudo echo" achieved by reshaping (b) to give a Gaussian envelope symmetrical about t=T.

The pseudo-echo spectra cannot be used for quantitative analysis since the intensity of signals in these spectra depend on both line-width and multiplet patterns of signals.

6. Pure Shift NMR

Homonuclear scalar coupling in ^{1}H NMR spectra provides great information to identify structure of organic molecules, inorganic molecules, natural products, and metabolites etc. However, it causes splitting of signals into multiplets which results in severely overlapped region of signals, leading to decrease resolution and complication in identification of number of signals. "Pure Shift" NMR spectra or "Chemical Shift" NMR spectra[56-64], is also known as broadband homonuclear decoupling which aims to merge multiplets into singlets to enhance resolution of spectra as well as simplification in identification of number of signals for their corresponding chemical shift which resembles proton decoupled ^{13}C-spectra in natural abundance without multiplets. Conventional and pure shift ^{1}H NMR spectra have been shown in Figure 1.13a) and 1.13b) respectively. There are a number of varieties of homonuclear broadband decoupling methods like 2D *J*-Resolved methods[44-49], methods based on time-reversal[65], bilinear rotational decoupling[66], constant time evolution methods[67], slice-selective (Zanger-Sterk) decoupling[54], and PSYCHE (pure shift yielded by chirp excitation) decoupling[68].

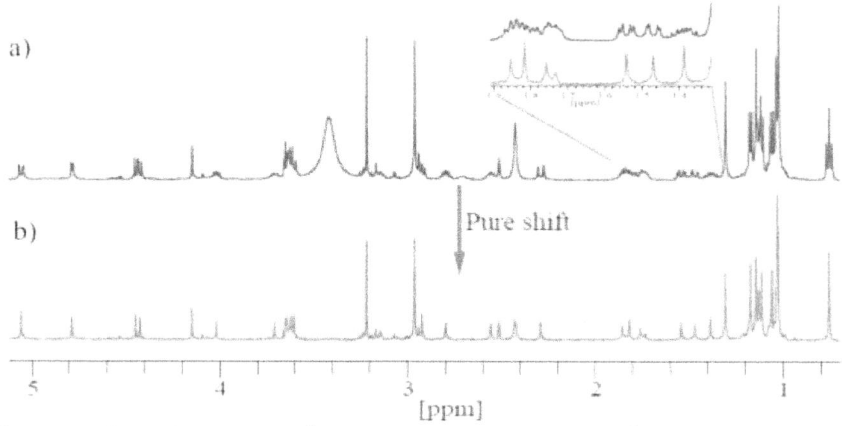

Figure 1.13. displays the conventional ^{1}H NMR spectrum in Fig. a), pure shift ^{1}H NMR spectra in Fig. b).

Pure shift NMR methods are applied such that the acquired time-domain signal shows average evolution only under the effect of chemical shift and not by *J*-coupling. The manipulation of the relative contribution of chemical shift and scalar coupling to the spin evolution is aimed by the pulse sequence. There are two main experimental approaches to achieve pure shift NMR spectra – (1) refocusing of scalar coupling in the evolution period (*J*-refocusing experiments), and (2) keeping constant the amount of evolution under the scalar coupling (constant-time).

6.1 Acquisition Modes to acquire spectra in absorption mode

6.1.1 Interferogram (Pseudo-2D)

Pseudo-2D NMR spectroscopy gives a suitable processing method for obtaining broadband homonuclear decoupled spectra which is valuable for structural identification of complex molecules. In any case, FID in such type of experiments are acquired as concatenated data points in the direct dimension over incremented time periods shown in Figure 1.14, resulting in long acquisition times without increment in sensitivity because of no averaging of signals between scans. Pseudo two-dimensional (2D) NMR have become particularly significant tool in producing broadband homodecoupled NMR spectra. It finds application in some experiments like and PSYCHE[68] pure shift, and the Zangger-Sterk (ZS) slice-selective[54]. These experiments are recognized from customary 2D homo decoupled experiments, for example, PSYCHE-TOCSY[69] , where the decoupling components are incorporated with the t_1 development periods offering ascend to a spectrum decoupled in the direct dimensionand subsequently no noteworthy time punishment is observed. In these techniques, decoupling in

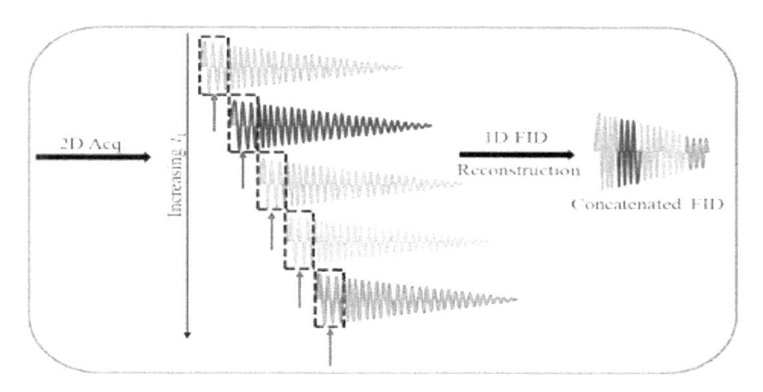

Figure 1.14. Conventional scheme for the acquisition of 1D homodecoupled ^{1}H NMR spectra. Interferogram (Pseudo-2D) acquisition mode that comprises on the recording of a 2D dataset followed by an FID remaking from the initial data chunks of each increment.

the direct dimension is accomplished by covariance NMR spectroscopy[70-74] a processing program that utilizes the covariance matrix of a series of 1D spectra to set up nuclear spin network according to 2D experiments. In pseudo-2D NMR experiments, the acquired FID is a one-dimensional interferogram produced by data points or data chunks achieved over discrete incremented time periods. Therefore, such methodologies take a long time to acquire acquisition comparable to that of a regular 2D NMR acquisition. However, this advancement exhibits significant resolution profit over "real-time" single scan methods[75-81] and also less prone to chunking artifact.

6.1.2 Real-time

Real-time homodecoupling method is used with many homodecoupling blocks to achieve homodecoupling in F_2 dimension. However, it works in a different manner than interferogram (pseudo-2D). In this method, homonuclear decoupled FID is directly recorded in a single scan with short sections like data chunks of FID which are separated by homodecoupling block in order to supress scalar coupling as shown in (Figure 1.15). This data chunk acquired in duration of τ period is derived as AQ/2n, where AQ denotes the acquisition time and n the number of loops. All acquired these data chunks are combined in a chain to construct an entire homondecoupled FID.

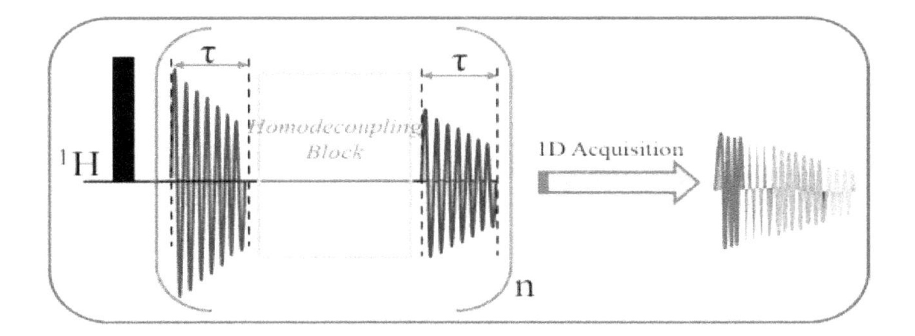

Figure 1.15. Real-time secure mode that incorporates connected FID chunks flanking the consolidation of the homodecoupling block[82, 25].

This method advantageously reduces the experimental time in comparison to the pseudo-2D method. However, this method has some demerits, depending on the involvement of inversion element: a) it has poor selectivity; b) the duration of data chunks are fixed to few tens in ms; c) ample loss of sensitivity on applying slice-selection; d) broader signals due to small value of T_2 relaxation due to which homodecoupling cannot be obtained efficiently; e) the interruption of the FID results in a discontinuity, causing the sidebands artefacts on both sides of the signals[81, 83]; f) inefficiency in decoupling of strongly coupled signals.

In summary, both real-time and pseudo-2D homodecoupling acquisition modes can be incorporated with the different homodecoupling blocks. These methods are better than all the other ps-NMR approaches and present specific advantages and disadvantages, additionally relying upon the molecules or spin systems for analysis.

Table 1.1

Acquisition Modes	Advantages	Disadvantages
Real-time	- Single scan. - Faster acquisition mode	- Broader signals. - Poorer selectivity. - Presence of artefacts. - Does not work with PSYCHE - More sensitivity than pseudo-2D.
Pseudo-2D	- Compatibility with any homodecoupling block.	- Longer acquisition time. - Special processing.

7. BIlinear Rotation Decoupling (BIRD)

The BIRD (bilinear rotation decoupling) scheme is a form of broadband homonuclear decoupling. Its pulse sequence element[84-96, 66] facilitates the selective inversion of ^{12}C-bound protons while remaining system in original state of the magnetization of protons bounded to ^{13}C. This is obtained by the scheme of pulse sequence shown in Fig. 1.16a, starting with 90° proton excitation, the magnetization of ^{13}C-bound protons evolve under only the scalar coupling to directly attached to ^{13}C.

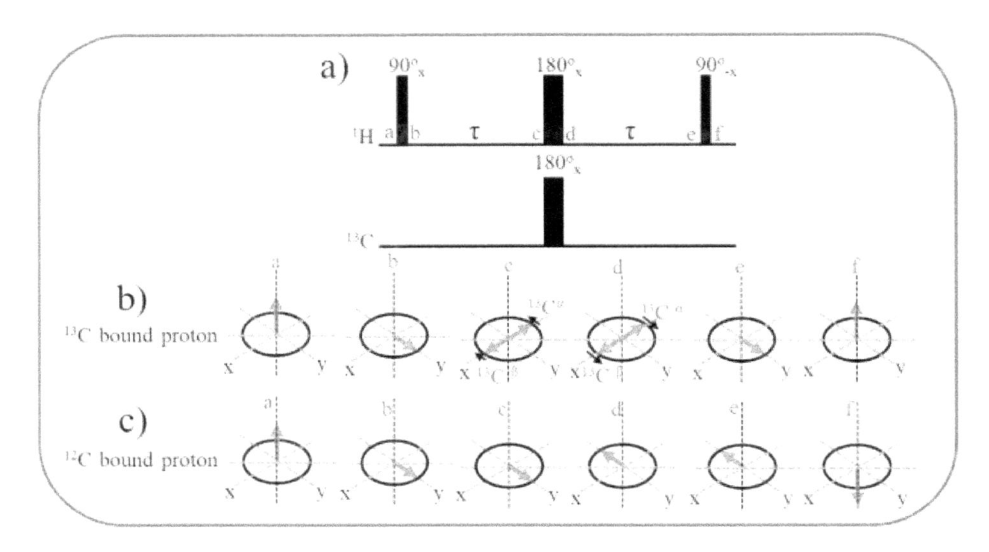

Figure 1.16. displays the pulse sequence of BIRD element ; (a) consists of narrow and wide filled black rectangles which are hard 90° and 180° pulses respectively with evolution period 2τ, and rotation of magnetizations of ^{13}C-bounded and ^{12}C- bounded protons along different axes in vector forms in b) and c), respectively.

After a delay τ of $1/^1J_{CH}$, the magnetizations of ^{13}C-bound proton become out of phase by 180° as compared to ^{12}C-bound protons. Chemical shift evolution is refocused by a hard 180° pulse in the middle of τ. The 90° pulse at the end of element rotates the magnetization of ^{13}C-bounded into the +z axis and ^{12}C-bounded protons onto -z-axis shown in Figure 1.16 b) and c) respectively. The decoupling between ^{13}C-bounded proton and its neighbours[97] can be achieved if BIRD element is placed in the middle of an evolution period t_1, which in a molecule at natural abundance are much unexpected to be bound to ^{13}C. Since this decoupling technique is applicable for ^{13}C-bound protons, its sensitivity is just around 1.1%

of a regular spectrum. However, this kind of decoupling provides entire inversion of passive spins (those that do not contribute to observed signal but to the couplings which cause multiplet pattern) while the remaining active spins (those that are bound to ^{13}C and do contribute to the signal) are unchanged, leading to high quality spectra in pure absorption mode. BIRD element cannot decouple geminal protons, since they both are bound to the same ^{13}C-atom. Therefore, this method is not able to distinguish both geminal protons. Other problems with this method occur in case of those compounds where larger differences in $^{1}J_{HC}$ values are found [98-99]. However, BIRD decoupling is advantageous as the parent proton spectra that show strong coupling mostly produce BIRD pure shift spectra without strong coupling artefacts. BIRD element is now being used as decoupling method in ^{1}H-^{13}C HSQC experiment for its pure shift spectra [100] at no additional cost of sensitivity as only ^{13}C-bound protons are observed through this method. The BIRD homonuclear decoupling method is used for achieving single-shot pure shift ^{1}H NMR spectra [101]. This decoupling method is also used in higher order of experiments like (3, 2)D BIRD^{r,x}-HSQC-TOCSY used for analysis of a complex carbohydrate mixture[102].

8. Pure Shift Yielded by Chirp Excitation (PSYCHE)

The PSYCHE (Pure Shift Yielded by Chirp Excitation)[68, 103-104] experiment is a useful tool in ^{1}H NMR spectroscopy for suppressing ^{1}H-^{1}H homonuclear J-coupling to attain homonuclear broadband decoupling proton spectra. This decoupling increases the resolution of spectra with a decrease in sensitivity which is less compared to BIRD and ZS approaches. The loss of the sensitivity of spectra usually occurs in a factor of 10 to 20 which depends on the experimental parameters. However, collapsing of multiplets into singlets produces a number of single signals at their single chemical shift, leading to enhancement in unambiguous identification of the number of proton signals. PSYCHE has a better spectral purity with ten folds sensitivity, and tolerance of strong coupling than other pure shift methods. However, It applies low flip (β) swept-frequency chirp pulses and a weak magnetic field gradient spontaneously shown in Fig. 1.17, avoids longer acquisition times, unable to deal efficiently with unwanted coherence transfer, strong coupling and inefficient data processing than other methods such as anti-z-cosy[105-106] which plays to its disadvantage. Both combined small flip β angle pulses refocus a small portion of spins (the active spins) in a stimulated echo, while leaving out the rest of the majorities of spins (passive spins). The overall effect of hard 180° pulse and combined chirp pulses along with weak gradient is to keep the minor spins (active spins) unchanged and to invert and dephase all the passive spins except the desired single quantum coherences of the active spins. Both spins are differentiated statistically, magnetization population of passive and active spins are proportional to $\sin^2\beta$ and $\cos^2\beta$ respectively.

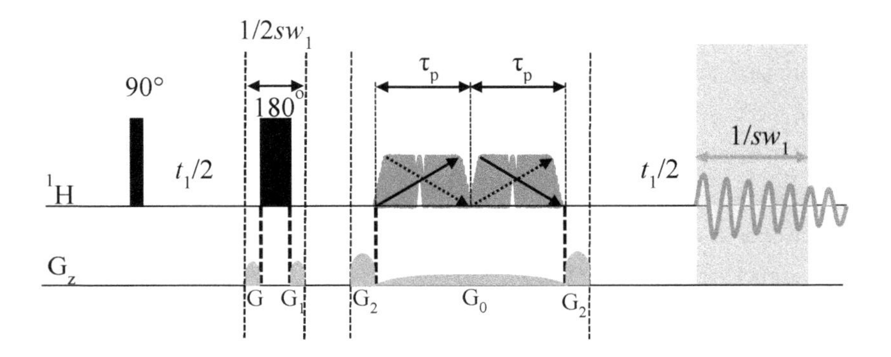

Figure 1.17. displays the pulse sequence for PSYCHE. Narrow filled rectangle is 90° pulse, broad filled rectangle is 180° pulse and trapezoids with one solid and second dotted diagonal arrow are both low-power frequency –swept chirp pulses of net flip angle $\beta << 90°$. Pulse with dotted, or solid arrows show the opposite directions of small flip angles which may be used simultaneously. G_0 is a weak pulse filed gradient with a sine shape; while G_1 and G_2 are also pulse field gradients with half sine shapes. The highlighted section in coloured rectangle of the FID with $1/SW_1$ represents the chunk of data acquired for each t_1 increment.

The cos function of the latter causes the intensity of desired signals from active spins to increase by increasing value of β but at a loss of spectral purity because of increasing artefacts in signals, as maximum purity of spectra belongs to minimum optimized value of small flip angle β [103]. It does not involve spatial encoding positions within sample like ZS method. Therefore, it has tolerance over strong coupling along with 10 fold higher sensitivity than ZS method involving spatial encoding where thick slices have very small quantity of magnetization populations of protons as compared to the whole sample and the BIRD method working on ^1H attached with very diluted ^{13}C atom.

^1H NMR spectrum and PSYCHE spectrum of β-estradiol molecules in solvent DMSO-d6 were acquired at 400 MHz spectrometer. Fig. 1.18(a) shows the molecular structure of β-estradiol, 1.18(b) for conventional 1D ^1H NMR spectrum, and 1.18c for PSYCHE spectrum. In 1.18(b), spectrum with overlapping of signals due to *J*-coupling leads to the ambiguity in assignments of signals that cause difficulty in assigning number of proton signals.

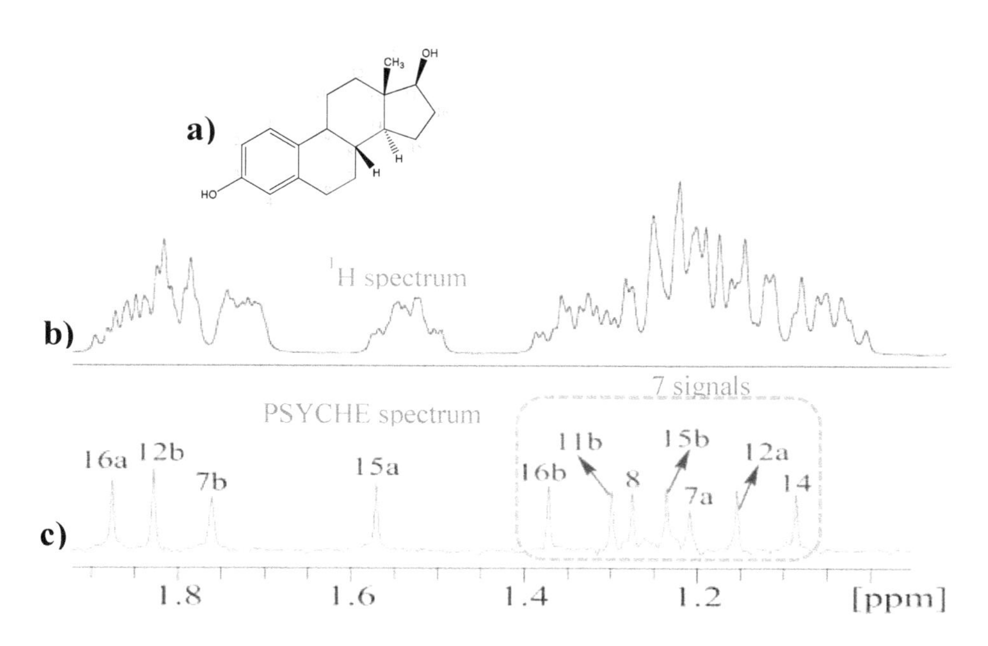

Figure 1.18. a) β-estradiol molecular structure, b) conventional 1D ^1H spectrum, and c) PSYCHE spectrum assigned signals. Both spectra were acquired in DMSO-*d6* solvent at 298 K.

However, Fig. 1.18c concludes the collapsing of multiplets into singlets by suppression of *J*-coupling by using PSYCHE technique which overcomes the difficulty to assign number of signals in the spectrum as the green colour box show the seven signals at their corresponding chemical shift. PSYCHE spectrum suffers from low sensitivity (~3-20 %) compared to conventional 1D ^1H NMR spectrum as that from 18b, the reduction factor is around 17. However, this technique is the most sensitive broadband homo-decoupling.

8.1 The Application of PSYCHE in Pure Shift Yielded by CHirp Excitation to DELiver Individual Couplings (PSYCHEDELIC)

PSYCHEDELIC is a homonuclear broadband decoupling method which extracts coupling constants from crowded ^{1}H spectra[107]. It is based on PSYCHE decoupling which was proposed by Sinnaeve et al. [108]. The PSYCHEDELIC method provides an approximate and tough method to measure coupling constants for structural, conformational, or stereochemical analysis[109]. The pulse sequence of PSYCHEDELIC method is shown in Fig. 1.19. It is very similar to that of the G-SERF experiment in principle of selective *J*-evolution during t_1.

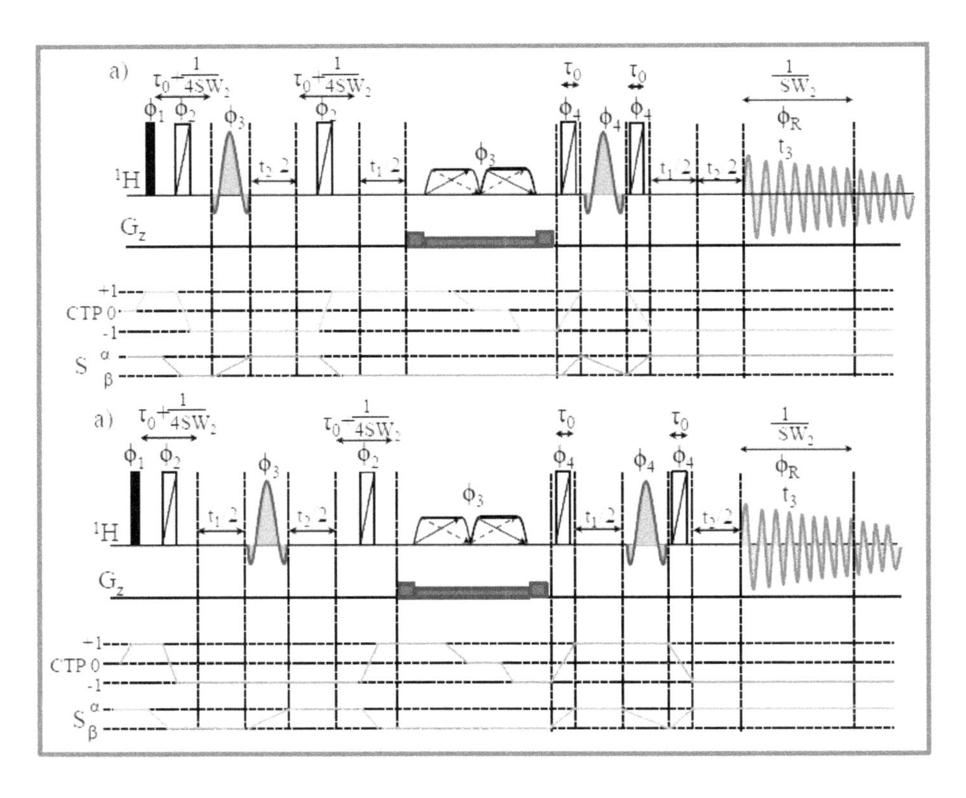

Figure. 1.19. displays the pulse sequence of PSYCHEDELIC experiment consists a) N-type and b) R-type. Narrow filled rectangle bars and trapezoids with double arrows represents 90° RF pulses and low-power chirp pulses of net flip angle β, sweeping frequency in a opposite directions simultaneously[104].Wide unfilled rectangles with diagonal lines are BIP720 180° pulse[113] and Black-filled half sine shape pulses are selective

180° pulses applied to the selected spin S. G_z indicates the pulse field gradients shown below line of pulse sequence. The chunk of data acquired for each increment in t_2 is shown by the highlighted part of the FID of duration 1/SW2. The selected coherence transfer pathway (CTP) and the evolution of the state of the selected spin S are shown.

However, it has two distinct key points from G-SERF. First, this method applies the PSYCHE decoupling element instead of a ZS element, because PSYCHE decoupling is far more efficient in achieving homonuclear broadband decoupling in crowded spectra. Second, to achieve absorption-mode spectra, PSYCHEDELIC utilizes the Pell-Keeler method (PK)[110] instead of a z-filter. The pulse sequence consists of combination of two sequences normal (N) and reversed (R) evolution in the same way as that of the classic echo/anti-echo processing[111], producing a 2D *J* spectrum with double absorption mode lineshapes. The PK method requires the state of selected spin during the direct acquisition time t_3 to be the same as it is during t_1 for both N- and R-type acquisition. In PSYCHEDELIC, the active couplings of selected spin (s) in t_1 needs to be preserved. The N- and R-type sequences are unlike the ones only in the t_1 evolution periods. The aim is to achieve suppression of unselected couplings in F_2. The selective pulses are used for the selective 2D *J* evolution and this period provides the evolution of chemical shift and selected couplings. For each increment in t_2, a chunk of data of duration $1/SW_2$ is achieved in t_3. The chunks are then composed, just like other interferogram-style pure shift experiments[112], to provide a 2D *J* data set in which the effects of all couplings are supressed except the selected couplings.

The molecule of β-estradiol has been used to record 2D PSYCHEDELIC spectrum as shown in Fig. 1.20(a). 1D ^1H NMR spectrum of that molecule has been shown in Fig. 1.20(b) and Fig. 1.20(c) shows 2D PSYCHEDELIC spectrum. The selective 180° pulse has been used to excite 9H signal which is selective for extracting coupling constants resulting from coupling with its surrounding signals. In Fig. 1.20a, 9H signals shows a correlation with its adjacent signals like 11α, 11β, and 8. The coupling constants were measured accurately by this method which provides the doublet of the coupling signals at their corresponding chemical shifts and disappearance of contour of that selected signal occurs while the diagonal peaks of the rest of signals appear at their corresponding chemical shift along F_2 and at zero chemical shift along F_1.

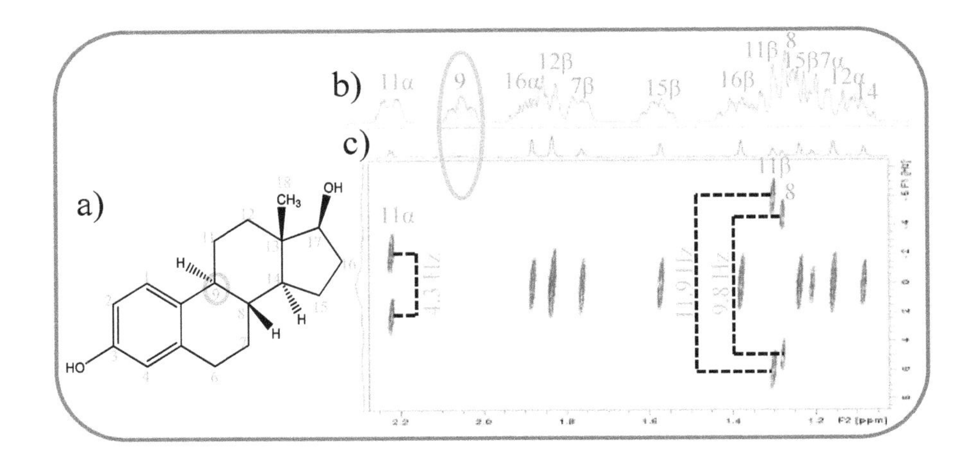

Figure. 1.20. (a) represents the β-estradiol molecular structure, (b) represents the extracted aliphatic region of 1D ^{1}H NMR spectrum with assigned signals of that molecule, and (c) represents the extracted section of 2D tilted 45° PSYCHEDELIC NMR spectrum with assigned coupling constants of sample β-estradiol molecule in solvent DMSO-*d6* at 400 MHz spectrometer. The selective 180° pulses were applied to proton 9H in this spectrum.

The 2D PSYCHEDELIC experiment assists pure shift spectrum to extract desired signals for measuring their coupling constants.

8.2 PSYCHE Based NMR Experiments

PSYCHE element has been associated with 1D and many 2D NMR experiments to acquire pure-shift spectra of 1D ^{1}H NMR[104], non-uniform sampling (NUS)[114-115], ^{1}H 2D TOCSY[116], RASA 2D *J* [117] to achieve absorption mode acquisition[118], and fast acquisition in a band-selective manner (BSE-PSYCHE-TOCSY)[119]. The most important experiments consisting of PSYCHE element are used for the measurement of desired individual scalar couplings in signals overlapped region by PSYCHEDELIC[107], and BSR-PSSE (Band-selective Refocusing Pure Shift Spin-Echo)[120]. There are others such as PSYCHE-EASY-ROESY[121] for direct observation of intermolecular NOEs in over-crowded spectrum, T_1 and

T_2 measurements by Inrec-PSYCHE[122], PSYCHE-CPMG-HSQMBC[123] for accurate measurement of heteronuclear couplings, and mixture analysis by PSYCHE-iDOSY [124].

9. Heteronuclear Multiple Quantum Coherence (HMQC)

Heteronuclear Multiple Quantum Coherence (HMQC)[125-126] is a two-dimensional inverse H,C correlation technique that determines the correlation between carbon (or other heteroatom) and directly attached protons through the heteronuclear multiple-quantum coherence transfer. The correlation between a heteronuclear *IS* spin system (I= ^{1}H, S= ^{13}C or ^{15}N), in which the *I* and *S* spins are directly bonded though covalent bond and the 1H spin I is homonuclear scalar coupled to a remote ^{1}H spin, *L*, results in homonuclear scalar coupling constant (J_{IL}) is considered as a much smaller than J_{IS}. The pulse sequence involved in HMQC experiment with heteronuclear spin system is shown in Fig. 1.21.

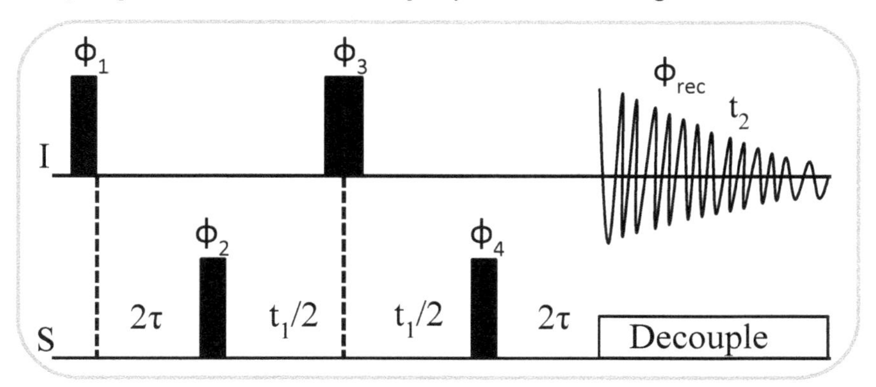

Figure 1.21. displays the pulse sequence for 2D HMQC experiment, filled narrow bars and wide bar are 90° and 180° pulses respectively. Decoupling during t_2 is obtained by applying WALTZ-16 and GARP-1 etc.

The product operators are generated in the pulse sequence are described during different time periods from initial to second ^{13}C 90° pulse.

$$I_z \xrightarrow{\pi/2(I_x + L_x) - 2\tau - \pi/2S_x} -2I_xS_y \quad (1)$$

$$\xrightarrow{t_1/2 - \pi(I_x + L_x) - t_1/2} -2I_xS_y \cos(\Omega_s t_1)\cos(\pi J_{IL} t_1) - 4I_yL_zS_y \cos(\Omega_s t_1)\sin(\pi J_{IL} t_1) \quad (2)$$

$$\xrightarrow{\pi/2S_x - 2\tau} -I_y \cos(\Omega_s t_1)\cos(\pi J_{IL} t_1) + 4I_xL_z \cos(\Omega_s t_1)\sin(\pi J_{IL} t_1) \quad (3)$$

The delay time 2τ is set to ½ J_{IS} in the pulse sequence and heteronuclear scalar coupling J_{IS} is evolved, while other homonuclear scalar couplings are ignored during 2τ time period.

Chemical shift evolution of the I spin is refocused by 180° (I) during the periods 2τ and t_1. The operator $2I_xS_y$ represents Heteronuclear multiple-quantum (MQ) coherence which does not evolve under the effect of the active scalar coupling, J_{IS}, during t_1 period. However, homonuclear J_{IL} scalar coupling evolves due to the effect of non-selective 180° (I) pulse upon both ^1H spins I and L. The resulting correlation between homonuclear spins in the spectrum exhibits homonuclear J coupling multiplet structure in the F_1 dimension. Finally, the F_2 lineshapes comprises of in-phase absorptive and anti-phase dispersive components, represented by I_y and $2I_xL_z$ operators in the third line, respectively.

Figure 1.22 a) represents the 2D conventional HMQC spectrum of glucose sample where crosspeaks are assigned to their corresponding nuclei.

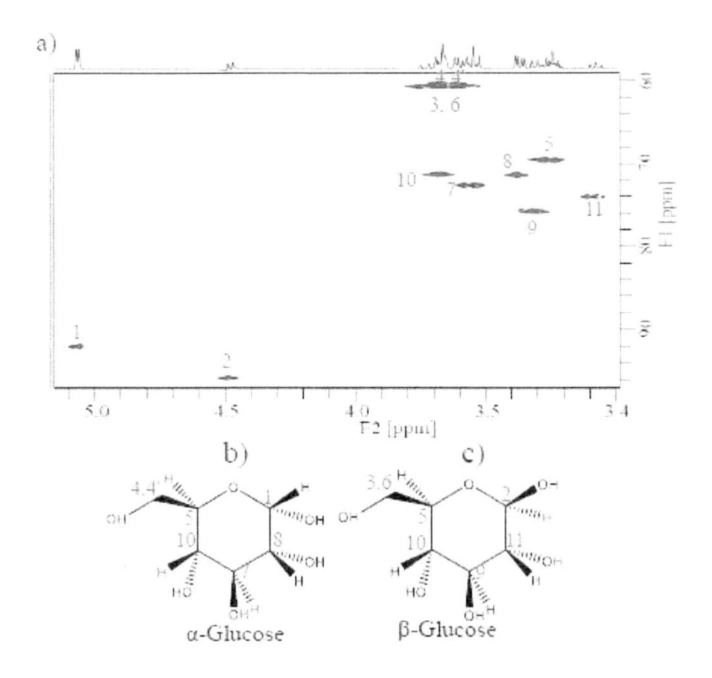

Figure 1.22. a) for 2D HMQC spectrum with assigned crosspeaks of glucose in D_2O solvent. The molecular structures of α-glucose and β-glucose are shown in Figures b) and c), respectively.

All the crosspeaks are associated with above mentioned anti-phase dispersive term which reduces resolution as well as the quality of the spectrum. In case of complex compounds where ^{13}C resonances are very closely resonating, this term causes the overcrowding of signals in the spectrum complicating the assignment of signals. Therefore, reliable techniques are needed to overcome such problem which can be addressed with our new technique discussed in chapter 4.

Reference:

[1] H. Kovacs, D. Moskau, M. Spral, Cryogenically cooled probes-a leap in NMR technology, *Prog. Nuc. Magn. Reson. Spec.*, 46(2005), 131-155.

[2] T. Helgaker, M. Jaszuński, P. Świder, Calculation of NMR Spin-Spin Coupling Constants in Strychnine. *Org. Chem.* 81 (2016) 11496–11500.

[3] D. Cremera, J. Gräfensteinb, Calculation and analysis of NMR spin–spin coupling constants. *PhysChemChem Phys*, 22 (2007) 2791-2816.

[4] V. M. R. Kakita, E. Kupče, J. Bharatam, Solid-state Hadamard NMR spectroscopy: Simultaneous measurements of multiple selective homonuclear scalar couplings, *J. Magn. Reson.* 251 (2015) 321-425.

[5] R. H. Contreras, J. E. Peralta, Angular dependence of spin-spin coupling constants, *Prog. Nucl. Magn. Reson. Spectrosc.* 37 (2000) 321-425.

[6] M. Karplus, Vicinal proton coupling in nuclear magnetic resonance, *J. Am. Chem. Soc.* 85 (1963) 2870-2871.

[7] W. P. Aue, J. Karhan, R. R. Ernst, Homonuclear broad band decoupling and two-dimensionnal *J*-resolved NMR spectroscopy *J. Chem. Phys.* 64 (1976) 4226-4227.

[8] T. Parella, Current development in homonuclear and heteronuclear *J*-resolved NMR experiments, *Magn. Reson. Chem.* 56 (2018) 230-250.

[9] T. Parella, J. F. Espinosa, Long-range proton-carbon coupling constants: NMR methods and applications, *Prog. Nucl. Magn. Reson Spectrosc.* 73 (2013) 17-55.

[10] D. Pitoux, Z. Hu, B. Plainchont, D. Merlet, J. Farjon, D. Bonnaffé, N. Giraud, Combining pure shift and J-edited spectroscopies: A stategy for extracting chemical shifts and scalar couplings from highly crowded proton spectra of oligomeric saccharided, *Magn. Reson. Chem.*, DOI: 10.1002/mrc. 4715 (2018).

[11] N. Nath, Lokesh, N. Suryaprakash, Measurement and application of long-range heteronuclear scalar couplings: Recent experimental and theoretical Developments, *ChemPhysChem* 13 (2012) 645-660.

[12] L. Castañar, T. Parella, Broadband ^{1}H homodecoupling NMR experiments: recent developments, methods and applications, *Magn Reson. Reson. Chem.* 53 (2015) 399-426.

[13] K. Zagger, Pure shift NMR, *Prog. Nucl. Magn. Reson. Spectrosc.* 86-87 (2015) 1-20.

[14] J. Jeener, Ampe`re Summer School, Basko Polje, Yugoslavia, 1971.

[15] W. P. Aue, E. Bartholdi, R. R. Ernst, Twodimensional spectroscopy. Application to nuclear magnetic resonance, *J. Chem. Phys.* 64 (1976) 2229–2246.

[16] J. Keeler, Understanding NMR spectroscopy pp. 190–191.

[17] A. A. Marchione, Two-dimensional NMR correlation experiments in the gas phase, *J. Magn. Reson.* 210 (2011) 31–37.

[18] M. K. Banjare, K. Behera, M. L. Satnami, S. Pandey and K. K. Ghosh, Host–guest complexation of ionic liquid with α- and β-cyclodextrins: a comparative study by ^{1}H-NMR, 13C-NMR and COSY, *New J. Chem.*, 42 (2018) 14542—14550.

[19] G. Dufour, B. Evrard, and P. de Tullio, 2D-Cosy NMR Spectroscopy as a Quantitative Tool in Biological Matrix: Application to Cyclodextrins, *American Association of Pharmaceutical Scientists*, 17 (2015) 1501-1510.

[20] F. Delaglio, Z. Wu, A. Bax, Measurement of Homonuclear Proton Couplings from Regular 2D COSY Spectra, *J. Magn. Reson.*, 149 (2001) 276–281.

[21] R. Brüschweiler, *J. C. Madsen, C. Griesinger, O. W.* Sørensen, *R. R. Ernst,* Two-Dimensional NMR Spectroscopy with Soft Pulses, *J. Magn. Reson.*, 73 (1987) 380-385.

[22] A. A. Maudsley, L. Muller and R. R. Ernst, Cross-correlation of spin-decoupled NMR spectra by heteronuclear two-dimensional spectroscopy, *J. Magn. Reson.*, 28 (1977) 463-469.

[23] C. Griesinger, O. W. Sørensen and R. R. Ernst, Practical aspects of the E.COSY technique. Measurement of scalar spin-spin coupling constants in peptides. *J. Magn. Reson.* 75 (1987) 474-492.

[24] K. Zangger, H. Sterk, Homonuclear Broadband-Decoupled NMR Spectra, *J. Magn. Reson.* 124 (1997) 486–489.

[25] N. H. Meyer and K. Zangger, Boosting the Resolution of 1H NMR Spectra by Homonuclear Broadband Decouple, *Chem. Phys. Chem.*, 15 (2014) 49–55.

[26] N. H. Meyer, K. Zangger, Enhancing the resolution of multi-dimensional heteronuclear NMR spectra of intrinsically disordered proteins by homonuclear broadband decoupling, *Chem. Commun.*, 50 (2014) 1488-1490.

[27] N. H. Meyer, K. Zangger, Viva la Resolution. Enhancing the resolution of ^{1}H-NMR spectra by broadband homonuclear decoupling, Synlett 25 (2014) 920–927.

[28] P. C. Lauterbur, Image formation by induced local interactions: examples employing nuclear magnetic resonance, Nature (1973) 190–191.

[29] A. Bax, R. Freeman, Enhanced NMR resolution by restricting the effective sample volume, *J. Magn. Reson.* 37 (1980) 177–181.

[30] L. Frydman, T. Scherf, A. Lupulescu, The acquisition of multidimensional NMR spectra within a single scan, *Proc. Natl. Acad. Sci. USA* 99 (2002) 15858–15862.

[31] N. M. Loening, J. Keeler, G. A. Morris, One-dimensional DOSY, *J. Magn. Reson.*15 (2001) 103–112.

[32] R. Freeman, The concept of spatially-encoded single-scan NMR, Concept Magn. Reson. 38A (2011) 1–6.

[33] M. J. Thrippleton, N. M. Loening, J. Keeler, A fast method for the measurement of diffusion coefficients: one-dimensional DOSY, *Magn. Reson. Chem.* 41 (2003) 441–447.

[34] N. M. Loening, M. J. Thrippleton, J. Keeler, R. G. Griffin, Single-scan longitudinal relaxation measurements in high-resolution NMR spectroscopy, *J. Magn. Reson.* 164 (2003) 321–328.

[35] H. T. Edzes, An analysis of the use of pulse multiplets in the single scan determination of spin–lattice relaxation rates, *J. Magn. Reson.* 17 (1975) 301–313.

[36] J. A. Aguilar, A.A. Colbourne, J. Cassani, M. Nilsson, G.A. Morris, Decoupling two-dimensional NMR spectroscopy in both dimensions, *Angew. Chem. Int. Ed.* 51 (2012) 6460–6463.

[37] J. J. Koivisto, Zero-quantum filtered pure shift TOCSY, *Chem. Commun.* 49 (2013) 96–98.

[38] G. A. Morris, J.A. Aguilar, R. Evans, S. Haiber, M. Nilsson, True chemical shiftcorrelation maps, *J. Am. Chem. Soc.* 132 (2010) 12770–12772.

[39] M. Foroozandeh, P. Giraudeau, D. Jeannerat, A toolbox of HSQC experiments for small molecules at high 13C-enrichment. Artifact-free, fully 13Chomodecoupled and JCC-encoding pulse sequences, Magn. Reson. Chem. 51 (2013) 808–814.

[40] M. Foroozandeh, P. Giraudeau, D. Jeannerat, Broadband 13C-homodecoupled heteronuclear single-quantum correlation nuclear magnetic resonance, *ChemPhysChem* 12 (2011) 2409–2411.

[41] A. J. Pell, J. Keeler, Two-dimensional *J*-spectra with absorption-mode lineshapes, *J. Magn. Reson.* 189 (2007) 293–299.

[42] N. Giraud, L. Beguin, J. Courtieu, D. Merlet, Nuclear magnetic resonance using a spatial frequency encoding: application to *J*-edited spectroscopy along the sample, *Angew. Chem. Int. Ed.* 49 (2010) 3481-3484.

[43] T. Fäcke, S. Berger, SERF, a new method for H, H spin-coupling measurement in organic chemistry, *J. Magn. Reson.,* Ser A 113 (1995) 114-116.

[44] W. P. Aue, J. Karhan, Richard R. Ernst, Homonuclear broad band decoupling and two-dimensional J-resolved NMR spectroscopy, J. Chem. Phys., 64(10) (1976), 4226-4227.

[45] Gareth A. Morris, Two-Dimensional *J*-Resolved Spectroscopy, eMagRes, (2009).

[46] R. Freeman G. Bodenhausen, R. Niedermeyer, David L. Turner, Double Fourier Transformation in High-Resolution NMR, J. Magn. Reson., 26 (1977), 133-164.

[47] Melcolm H. Levitt, Spin Dynamics, John Wiley & Sons Ltd., Second ed. (2008).

[48] J. Keeler, A. J. Shaka, R. Freeman, Separation of Chemical Shifts and Spin Coupling in Proton NMR. Elimination of Dispersion Signals from Two-Dimensional Spectra, J. Magn. Reson., 56 (1984), 294-313.

[49] James W. Emsley, Gareth A. Morris, Multidimensional NMR Methods for the Solution State, John Wiley & Sons Ltd., (2010).

[50] E. L. Hahn, Spin Echoes, Phys. Rev., 80(4) (1950), 580

[51] R. Freeman, G. Bodenhausen, Gareth A. Morris, David L. Turner, NMR spectra of some simple spin systems studied by two-dimensional fourier transformation of spin echoes, J. Magn. Reson., 31(1) (1978), 75-95.

[52] J. Keeler, D. Neuhaus, Comparison and evaluation of methods for two-dimensional NMR spectra with absorption-mode lineshapes, J. Magn. Reson. 63 (1985) 454–472.

[53] P. Bachmann, W.P. Aue, L. Mu¨ller, R.R. Ernst, Phase separation in two-dimensional spectroscopy, J. Magn. Reson. 28 (1977) 29–39.

[54] K. Zangger, H. Sterk, Homonuclear Broadband-Decoupled NMR Spectra, J. Magn. Reson. 124 (1997) 486-489.

[55] A. Bax, R. Freeman, G. A. Morris, A Simple Method for Suppressing Dispersion-Mode Contributions in NMR Spectra: The "Pseudo Echo", J. Magn. Reson. 43 (1981) 333-338.

[56] R. W. Adams, Pure Shift NMR Spectroscopy, *eMagRes*. 3 (2014) 295-309.

[57] J. A. Aguilar, S. Faulkner, M. Nilsson and G.A. Morris, Pure Shift ^{1}H NMR: A Resolution of the Resolution Problem, *Angew. Chem. Int. Ed.;* 2010, 49, 3901-3903.

[58] M. Foroozandeh, R.W. Adams, M. Nilsson, G.A. Morris, Ultrahigh-resolution NMR spectroscopy, Angew. Chem. Int. Ed. 53 (2014) 6990–6992.

[59] M. Foroozandeh, R.W. Adams, M. Nilsson, G.A. Morris, Ultrahigh-resolution total correlation NMR spectroscopy, J. Am. Chem. Soc. 136 (2014) 11867–11869.

[60] V.M.R. Kakita, K. Rachineni, R.V. Hosur, Fast and simultaneous determination of ^{1}H–^{1}H and ^{1}H–^{19}F scalar couplings in complex spin systems: Application of PSYCHE homonuclear broadband decoupling, Magn. Reson. Chem. 56 (2018) 1043–1046.

[61] V.M.R. Kakita, R.V. Hosur, Non-uniform-sampling ultrahigh resolution TOCSY NMR: analysis of complex mixtures at microgram levels, ChemPhysChem 17 (2016) 2304–2308.

[62] L. Kaltschnee, K. Knoll, V. Schmidts, R.W. Adams, M. Nilsson, G.A. Morris, C.M. Thiele, Extraction of distance restraints from pure shift NOE experiments, *J. Magn. Reson.* 271 (2016) 99–109.

[63] J.A. Aguilar, R. Belda, B.R. Gaunt, A.M. Kenwright, I. Kuprov, Separating the coherence transfer from chemical shift evolution in high-resolution pure shift COSY NMR, Magn Reson Chem. 56 (2018) 969–975.

[64] K. Zanger. Progress NMR Spectroscopy 2015, 86-87, 1-20.

[65] C. Griesinger, O.W. Soerensen, R.R. Ernst, Two-dimensional correlation of connected NMR transitions, J. Am. Chem. Soc. 107 (1985) 6394–26396.

[66] J. R. Garbow, D. P. Weitekamp, A. Pines, Bilinear rotation decoupling of homonuclear scalar interactions, Chem. Phys. Lett. 93 (1982) 504–509.

[67] A. Bax, A.F. Mehlkopf, J. Smidt, Homonuclear broadband-decoupled absorption spectra, with linewidths which are independent of the transverse relaxation rate, J. Magn. Reson. 35 (1979) 167–169.

[68] M. Foroozandeh, G. A. Morris, and M. Nilsson, PSYCHE Pure Shift NMR Spectroscopy, *Chem. Eur. J.* 24 (2018), 13988 – 14000.

[69] M. Foroozandeh, R. W. Adams, M. Nilsson, G. A. Morris, Ultrahigh-Resolution Total Correlation NMR Spectroscopy, *J. Am. Chem. Soc.* 136 (2014) 11867–11869.

[70] A. Fredi, P. Nolis, C. Cobas, G. E. Martin, T. Parella, Exploring the use of Generalized Indirect Covariance to reconstruct pure shift NMR spectra: Current Pros and Cons, *J. Magn. Reson.* 266 (2016) 16–22.

[71] A. Fredia, P. Nolis, C. Cobas, T. Parella, Access to experimentally infeasible spectra by pure-shift NMR covariance, *J. Magn. Reson.*270 (2016) 161-168.

[72] M. Jaeger, R. L. E. G. Aspers in Annu. Reports NMR Spectrosc. (Ed.: G. A. Webb), Academic Press, Oxford, 2014, pp. 271–349.

[73] F. Zhang, R. Breschweiler, Indirect Covariance NMR Spectroscopy, J. Am. Chem. Soc. 126 (2004) 13180–13181.

[74] R. Brüschweiler, F. Zhang, Covariance nuclear magnetic resonance spectroscopy, *J. Chem. Phys.* 120 (2004) 5253–5260.

[75] I. E. Ndukwe, C. P. Butts, Pure-shift IMPRESS EXSIDE – Easy measurement of 1H–13C scalar coupling constants with increased sensitivity and resolution, RSC Adv. 2015, 5, 107829–107832.

[76] L. CastaÇar, P. Nolis, A. Virgili, T. Parella, Full Sensitivity and Enhanced Resolution in Homodecoupled Band-Selective NMR Experiments, *Chem. Eur. J.* 2013, 19, 17283 – 17286.

[77] N. H. Meyer, K. Zangger, Simplifying Proton NMR Spectra by Instant Homonuclear Broadband Decoupling, Angew. Chem. Int. Ed. 52 (2013) 7143–7146.

[78] N. H. Meyer, K. Zangger, Boosting the Resolution of ^{1}H NMR Spectra by Homonuclear Broadband Decoupling, *ChemPhysChem* 15 (2014) 49–55.

[79] R. W. Adams, L. Byrne, P. Király, M. Foroozandeh, L. Paudel, M. Nilsson, J. Clayden, G. A. Morris, Diastereomeric ratio determination by high sensitivity band-selective pure shift NMR
spectroscopy, Chem. Commun. 50 (2014) 2512–2514.

[80] L. CastaÇar, J. Saur, P. Nolis, A. Virgili, T. Parella, Implementing homo- and heterodecoupling in region-selective HSQMBC Experiments, *J. Magn. Reson.* 283 (2014) 63–69.

[81] L. Paudel, R. W. Adams, P. Király, J. A. Aguilar, M. Foroozandeh, M. J. Cliff, M. Nilsson, P. Sándor, J. P. Waltho, G. A. Morris, Simultaneously Enhancing Spectral Resolution and Sensitivity in Heteronuclear Correlation NMR Spectroscopy, Angew. Chem. Int. Ed. 2013, 52, 11616–11619.

[82] L. Castañar, T. Parella, Broadband ^{1}H homodecoupled NMR experiments: recent developments, methods and applications. Magn. Reson. Chem. 53(2015) 399–426.

[83] K. Zangger, Pure shift NMR. Prog. Nucl. Magn. Reson. Spectrosc. 86–87, 1–20 (2015).

[84] A. Bax, Broadband homonuclear decoupling in heteronuclear shift correlation NMR spectroscopy, J. Magn. Reson. 53 (1983) 517–520.

[85] D. Uhrín, T. Liptaj, K.E. Kövér, Modified BIRD pulses and design of heteronuclear pulse sequences, J. Magn. Reson. 101 (1993) 41–46.

[86] J. A. Aguilar, M. Nilsson, G. A. Morris, Simple proton spectra from complex spin systems: pure shift NMR spectroscopy using BIRD, Angew. Chem. Int. Ed. 50 (2011) 9716–9717.

[87] T. Reinsperger, B. Luy, Homonuclear BIRD-decoupled spectra for measuring one-bond couplings with highest resolution: CLIP/CLAP-RESET and constant time-CLIP/CLAP-RESET, J. Magn. Reson. 239 (2014) 110–120.

[88] A. Verma, B. Baishya, Real-time bilinear rotation decoupling in absorptive mode *J*-spectroscopy: detecting low-intensity metabolite peak close to high intensity metabolite peak with convenience, J. Magn. Reson. 266 (2016) 51–58.

[89] L. Kaltschnee, A. Kolmer, I. Timári, V. Schmidts, R.W. Adams, M. Nilsson, K.E. Kövér, G.A. Morris, C.M. Thiele, "Perfecting" pure shift HSQC: full homodecoupling for accurate and precise determination of heteronuclear couplings, Chem. Commun. 50 (2014) 15702–15705.

[90] A. Lupulescu, G. L. Olsen, L. Frydman, Toward single-shot pure-shift solution 1H NMR by trains of BIRD-based homonuclear decoupling, J. Magn. Reson. 218 (2012) 141–146.

[91] L. Paudel, R.W. Adams, P. Király, J. A. Aguilar, M. Foroozandeh, M.J. Cliff, M. Nilsson, P. Sándor, J. P. Waltho, G. A. Morris, Simultaneously enhancing spectral resolution and sensitivity in heteronuclear correlation NMR spectroscopy, Angew. Chem. Int. Ed. 52 (2013) 11616–11619.

[92] I. Timari, C. Wang, A.L. Hansen, G.C. dos Santos, S.O. Yoon, L. Bruschweiler-Li, R. Brüschweiler, Real-time pure shift HSQC NMR for untargeted metabolomics, Anal. Chem. 91 (2019) 2304–2311.

[93] A. Verma, R. Parihar, S. Bhattacharya , B. Baishya, Analyses of complex mixtures by F_1 homo-decoupled diagonal suppressed total correlation spectroscopy, 18 (2017) 3076–3082.

[94] J.D. Haller, A. Bodor, B. Luy a Real-time pure shift measurements for uniformly isotope-labeled molecules using X-selective BIRD homonuclear decoupling, J. Magn. Reson. 302 (2019) 64–71

[95] P. Nolis, K.M. Corral, M.P. Trujillo, T. Parella, Broadband homodecoupled timeshared 1H–13C and 1H–15N HSQC experiments, J. Magn. Reson. 298 (2019) 23–30.

[96] P. Sakhaii, B. Haase, W. Bermel, Experimental access to HSQC spectra decoupled in all frequency dimensions, J. Magn. Reson. 199 (2009) 192–198.

[97] A. Bax, Broadband homonuclear decoupling in heteronuclear shift correlation NMR spectroscopy, *J. Magn. Reson.* 53 (1983) 517–520.

[98] G. Otting, K. Wüthrich, Extended heteronuclear editing of 2D 1H NMR spectra of isotope-labeled proteins, using the X(ω_1, ω_2) double half filter, *J. Magn. Reson.* 85 (1989) 586–594.

[99] C. Zwahlen, P. Legault, Sébastien J.F. Vincent, J. Greenblatt, R. Konrat, L.E. Kay, Methods for measurement of intermolecular NOEs by multinuclear NMR spectroscopy: application to a bacteriophage k N-peptide/boxB RNA complex, *J. Am. Chem. Soc.* 119 (1997) 6711–6721.

[100] P. Kiraly, M. Nilsson, G. A. Morris, Practical aspects of real-time pure shift HSQC experiments, *Magn Reson Chem.* 56 (2018) 993–1005.

[101] A. Lupulescu, G. L. Olsen, L. Frydman, Toward single-shot pure-shift solution 1H NMR by trains of BIRD-based homonuclear decoupling, *J. Mag. Reson.* 218 (2012) 141–146.

[102] N. Brodaczewska, Z. Košťálová, D. Uhrín, (3, 2) D ^{1}H, ^{13}C BIRD^{r,x} -HSQC-TOCSY for NMR structure elucidation of mixtures: application to complex carbohydrates, *J. Biomol.* NMR 70 (2018) 115–122.

[103] M. Foroozandeh, R. W. Adams, N. J. Meharry, D. Jeannerat, M. Nilsson, G. A. Morris, Ultrahigh-Resolution NMR Spectroscopy, *Angew. Chem. Int. Ed.* 53 (2014) 6990 –6992.

[104] M. Foroozandeh, Ralph W. Adams, P. Kiraly, M. Nilsson, Gareth A. Morris, Measuring couplings in crowded NMR spectra: pure shift NMR with multiplet analysis, *Chem. Commun.*, 51(84) (2015), 15410-15413.

[105] A. J. Pell, R. A. E. Edden, J. Keeler, Broadband proton-decoupled proton spectra, *Magn. Reson. Chem.* 45 (2007) 45 296 – 316.

[106] H. Oschkinat, A. Pastore, P. Pfndler, G. Bodenhausen, *J. Magn. Reson.* 69 (1986) 559-566.

[107] M. Foroozandeh, D. Sinnaeve, M. Nilsson, Gareth A. Morris, A General Method for Extracting Individual Coupling Constants from Crowded 1H NMR Spectra, *Angew. Chem. Int. Ed.*, 55 (2016), 1090-1093.

[108] D. Sinnaeve, M. Foroozandeh, M. Nilsson, G.A. Morris, A general method for extracting individual coupling constants from crowded 1H NMR spectra, Angew. Chem. Int. Ed. 55 (2016) 1090-1093.

[109] D. Sinnaeve, M. Foroozandeh, M. Nilsson, and G. A. Morris, A General Method for Extracting Individual Coupling Constants from Crowded ^{1}H NMR Spectra, *Angew. chem. Int. Ed.*, 55 (2016) 1090-1093.

[110] A.J. Pell, J. Keeler, Two-dimensional J-spectra with absorption-mode lineshapes, J. Magn. Reson. 189 (2007) 293-299.

[111] J. Keeler, D. Neuhaus, comparison and evolution of methods for two-dimensional NMR spectra with absorption-mode lineshapes, *J. Magn. Reson.* 1985, 63, 454-.

[112] a) B. Bçttcher, C. M. Thiele in eMagRes, Vol. 1 (Ed.: R.Wasylishen), Wiley, Chichester, 2012, pp. 169–180, DOI:10.1002/9780470034590.emrstm1194; b) G. Kummerlçwe, B. Luy, TrAC Trends Anal. Chem. 2009, 28, 483; c) W. A. Thomas, Prog. Nucl. Magn. Reson. Spectrosc. 1997, 30, 183.

[113] Mari A. Smith, Haitao Hu,1 and A. J. Shaka, Improved Broadband Inversion Performance for NMR in Liquids *J. Magn. Reson.* 151 (2001) 269-283.

[114] Veera M. R. Kakita, Ramakrishna V. Hosur, Non-Uniform-Sampling Ultrahigh Resolution TOCSY NMR: Analysis of Complex Mixtures at Microgram Levels, *Chemphyschem*, 17(15) (2016), 2304-2308.

[115] I. E. Ndukwe, A. Shchukina, V. Zorin, C. Cobas, K. Kazimierczuk, C. P. Butts, Enabling Fast Pseudo-2D NMR Spectral Acquisition for Broadband Homonuclear Decoupling: The EXACT NMR Approach, *Chemphyschem* 18(15), (2017), 2081-2087.

[116] M. Foroozandeh, Ralph W. Adams, M. Nilsson, Gareth A. Morris, Ultrahigh-resolution total correlation NMR spectroscopy, *J. Am. Chem. Soc.*, 136(34) (2014), 11867-11869.

[117] Y. Huang, Y. Yang, S. Cai, Z. Chen, H. Zhan, C. Li, C. Tan, Z. Chen, General Two-Dimensional Absorption-Mode J-Resolved NMR Spectroscopy, *Anal. Chem.*, 89(23) (2017), 12646-12651.

[118] P. Kiraly, M. Foroozandeh, M. Nilsson, Gareth A. Morris, Anatomising proton NMR spectra with pure shift 2D *J*-spectroscopy: A cautionary tale, *Chem. Phys. Lett.*, 683 (2017), 398-403.

[119] Veera M. R. Kakita, S. P. Vemulapalli, J. Bharatam, Band-selective excited ultrahigh resolution PSYCHE-TOCSY: fast screening of organic molecules and complex mixtures, *Magn. Reson. Chem.*, 54(4) (2016), 308-314.

[120] Veera M. R. Kakita, Kanaka M. Jerripothula, Sahitya P. B. Vemulapalli, J. Bharatam, Selective measurement of ^{1}H-^{1}H scalar couplings from crowded chemical shift regions: Combined pure shift and spin-echo modulation approach, *Magn. Reson. Chem.*, 56 (10) (2018), 941-946.

[121] Eliska Prochzkov, A. Kolmer, J. Ilgen, M. Schwab, L. Kaltschnee, M. Fredersdorf, V. Schmidts, R. C. Wende, Peter R. Schreiner, Christina M. Thiele, Uncovering Key Structural Features of an Enantioselective Peptide-Catalyzed Acylation Utilizing Advanced NMR Techniques, *Angew. chem. Int. Ed.*, 55 (2016), 15754-15759.

[122] Veera M. Rao Kakita, V. K. Shukla, M. Bopardikar, T. Bhattacharya, Ramakrishna V. Hosur, Measurement of ^{1}H NMR relaxation times in complex organic chemical systems: application of PSYCHE, *RSC Adv.*, 6(102) (2016), 100098-100102.

[123] I. Timari, L. Szilagyi, Katalin E. Kover, PSYCHE CPMG-HSQMBC: An NMR Spectroscopic Method for Precise and Simple Measurement of Long-Range Heteronuclear Coupling Constants, *Chemistry-A Eur. J.*, 21(40) (2015) ,13939-13942.

[124] M. Foroozandeh, L. Castanar, L. G. Martins, D. Sinnaeve, G. D. Poggetto, C. F. Tormena, Ralph W. Adams, Gareth A. Morris, M. Nilsson, Ultrahigh-Resolution Diffusion-Ordered Spectroscopy, *Angew. Chem. Int. Ed.*, 55(50) (2016), 15579-15582.

[125] L. Muller, Sensitivity enhanced detection of weak nuclei using heteronuclear multiple quantum coherence. *J Am Chem Soc* 101 (1979) 4481– 4484.

[126] A. Bax, R. Griffey, B. Hawkins, Correlation of proton and nitrogen-15 chemical shifts by multiple quantum NMR. *J Magn Reson* 55, (1983) 301–315.

Chapter 2

Parallel acquisition of slice-selective ^{1}H-^{1}H soft COSY spectra

1. Introduction

Quest for high resolution has given rise to a large number of novel pulse techniques in recent years for simplifying analysis of complex proton NMR spectra. Among these various schemes one notable approach is slice-selective broad band ^{1}H–^{1}H decoupling utilizing the Zangger and Sterk (ZS) pulse sequence element, which allows pure chemical shift evolution to be spatially encoded along the NMR sample finally giving rise to a singlet only spectrum free from otherwise complex multiplet peaks[1–9]. Such an approach provides access to chemical shift information in overcrowded spectral regions. Various 2D experiments have been upgraded with the high-resolution offered by this technique[10–13].

The other methodology that stemmed from slice selective encoding is the *J*-edited spectroscopy or gradient-encoded selective refocusing experiment (G-SERF) that focused mainly on obtaining the other crucial parameter the *J*-coupling information[14–17]. This approach allows assignment and measurement of all the homonuclear scalar couplings for a given proton site through parallel acquisition of all the SERF experiment (which are spatially encoded) related to that site.

Slice-selective decoupling employs simultaneous application of semi-selective pulse and pulsed field gradient along the axis of the NMR tube. The presence of a linear gradient introduces different magnetic field strengths in different parts of the NMR sample. As a result, the resonance frequency of different resonances undergoes a location dependent frequency shift over the length of the active sample. This process is also termed as spatial frequency encoding. The original ZS slice-selective broadband decoupling scheme utilized a soft and a hard 180° pulse sequentially during a weak field gradient pulse. The soft 180° pulse selectively inverts different signals in different slices of the sample. Thus, no other signal except a particular signal experiences the refocusing pulse in a chosen slice achieving the decoupling from other spins in that slice. Chemical shift evolution can then be carried out for that signal if followed by a hard refocusing pulse which affect all the spins in the same slice. This arrangement gives rise to the broadband decoupling of homonuclear scalar coupling interactions while ensuring a pure chemical shift evolution for all the signals. In *J*-Edited spectroscopy at the centre of t_1 evolution period the soft 180° pulse (in the presence of gradient) is followed by another soft 180° pulse (in the absence of gradient) on one of the signals. This accomplishes spatially encoded evolution of all the scalar couplings that involve that specific signal on which the later soft 180° pulse has been applied. Each scalar coupling interaction evolves in a different slice so that in principle no two couplings evolve in the same slice. However, chemical shift information is in general lost in the t_1 dimension in such an encoding scheme.

In the present work, we demonstrate a different aspect of the spatial frequency encoding which can give rise to slice selective (or spatially encoded) mixing and bring about a simplification of all the correlation peaks between a given proton site and all its coupled partners in a single experiment. In other words, such a slice selective mixing strategy allows parallel acquisition of several soft COSY spectra. Each of the soft COSY spectra between different pairs of spins is encoded in a different slice. In favourable cases this can be accomplished for multiple proton sites (which are not coupled) simultaneously scaling up the number of experiments that can be acquired parallelly. The new experiment is coined as PASS-soft-COSY (Parallelly Acquired Slice-Selective Soft COSY). There exists other ways of parallel data acquisition schemes in liq- uid state NMR, most notable single scan (ultra-fast) spectroscopy[18,19], homonuclear broadband decoupling[1], relaxation-delay free data acquisition[4], G-SERF[14], multiple FID acquisition[20], and simultaneously cycled NMR spectroscopy[21].

There exists many other ways of simplifying complex proton spectra such as SERF[22] (Selective Refocusing), soft-COSY, and Z-COSY techniques[23–26]. SERF is a selective refocusing method where scalar coupling between a selected pair of spins are allowed to evolve during t_1 by applying selective refocusing pulses on both spins at the centre of t_1 dimension yielding a selective *J*-resolved spectrum. The simplified spectrum retains splitting between two selected spins only. However, the method demands $n(n-1)/2$ two-dimensional experiments to extract all the coupling constants in a system of n coupled signals. The G-SERF or *J*-edited spectroscopy allows parallel acquisition of all the SERF experiment for a given signal reducing the number of separate 2D experiments that are required to (n-1). The axial peak artifact in G-SERF experiment has been overcome by the recent clean G-SERF experiment[17]. Akin to G-SERF, more recently, the PSYCHEDELIC experiment has been demonstrated to produce couplings between chosen spins along F_1 but with pure chemical shift along F_2 and with resolution close to natural line width[29].

The soft COSY and Z-COSY type of experiments improve resolution by utilizing selective pulses or small flip angle pulses during mixing period[23–26]. A reduced multiplet pattern in soft COSY type of experiments contains minimum number of transitions and yet display all the coupling information present which can get obscured in regular COSY or other experiments that uses non- selective pulses for mixing. Extraction of the coupling constants from the resulting simplified and well resolved spectrum is less tedious. Although selective pulses are used in the G-SERF and SERF experiments also, it is the t_1 evolution that is basically manipulated by selective pulses to retain limited number of coupling interactions, and refocusing experiments do not generally have any mixing period. Herein, we focus on generating simpler (well resolved) spectra utilizing the soft COSY type of selective coherence transfer but utilizing a 90° ZS pulse sequence element for the mixing instead of regular selective pulses.

2 Experimental Section

2.1 Product Operator calculations in the pulse sequence of PASS soft COSY

A three spin system with homonuclear scalar couplings between all the spins are considered.

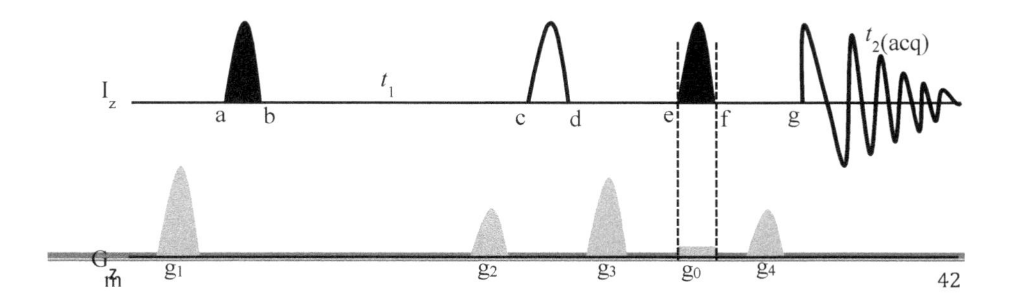

$$\underline{\text{At a}}\quad \begin{matrix} I_{2z} \\ I_{3z} \\ I_{1z} \end{matrix} \xrightarrow{\pi/2I_{2x}} \underline{\text{At b}}\quad \begin{matrix} -I_{2y} \\ I_{3z} \\ I_{1z} \end{matrix} \xrightarrow{\Omega_2 t_1 I_{2z}} -I_{2y}\cos(\Omega_2 t_1) + I_{2x}\sin(\Omega_2 t_1)$$

$$\xrightarrow{2\pi j_{23} t_1 I_{2z} I_{3z}} -I_{2y}\cos(\Omega_2 t_1)\cos(\pi j_{23} t_1) + 2I_{2x}I_{3z}\cos(\Omega_2 t_1)\sin(\pi j_{23} t_1) + I_{2x}\sin(\Omega_2 t_1)\cos(\pi j_{23} t_1) + 2I_{2y}I_{3z}\sin(\Omega_2 t_1)\sin(\pi j_{23} t_1) \quad \text{-------(1)}$$

Thus for each homonuclear scalar couplings four terms are obtained.

$$\xrightarrow{2\pi j_{21} t_1 I_{2z} I_{1z}}$$

$$- I_{2y}\cos(\Omega_2 t_1)\cos(\pi j_{23} t_1)\cos(\pi j_{21} t_1) + 2I_{2x}I_{1z}\cos(\Omega_2 t_1)\cos(\pi j_{23} t_1)\sin(\pi j_{21} t_1)$$

$$+ 2I_{2x}I_{3z}\cos(\Omega_2 t_1)\sin(\pi j_{23} t_1)\cos(\pi J_{21} t_1) + 4I_{2y}I_{3z}I_{1z}\cos(\Omega_2 t_1)\sin(\pi j_{23} t_1)\sin(\pi j_{21} t_1)$$

$$+ I_{2x}\sin(\Omega_2 t_1)\cos(\pi j_{23} t_1)\cos(\pi j_{21} t_1) + 2I_{2y}I_{1z}\sin(\Omega_2 t_1)\cos(\pi j_{23} t_1)\sin(\pi j_{21} t_1)$$

$$+ 2I_{2y}I_{3z}\sin(\Omega_2 t_1)\sin(\pi j_{23} t_1)\cos(\pi j_{21} t_1) - 4I_{2x}I_{3z}I_{1z}\sin(\Omega_2 t_1)\sin(\pi j_{23} t_1)\sin(\pi j_{21} t_1) \quad \text{-------(2)}$$

$$\xrightarrow{\pi/2I_{2x}\ \text{(between time point 'c' and 'd')}}$$

$$+ I_{2z}\cos(\Omega_2 t_1)\cos(\pi j_{23} t_1)\cos(\pi j_{21} t_1) + 2I_{2x}I_{1z}\cos(\Omega_2 t_1)\cos(\pi j_{23} t_1)\sin(\pi j_{21} t_1) + 2I_{2x}I_{3z}\cos(\Omega_2 t_1)\sin(\pi j_{23} t_1)\cos(\pi j_{21} t_1) - 4I_{2z}I_{3z}I_{1z}\cos(\Omega_2 t_1)\sin(\pi j_{23} t_1)\sin(\pi j_{21} t_1) +$$

$$I_{2x}\sin(\Omega_2 t_1)\cos(\pi j_{23} t_1)\cos(\pi j_{21} t_1) - 2I_{2z}I_{1z}\sin(\Omega_2 t_1)\cos(\pi j_{23} t_1)\sin(\pi j_{21} t_1) - 2I_{2z}I_{3z}\sin(\Omega_2 t_1)\sin(\pi j_{23} t_1)\cos(\pi j_{21} t_1) - 4I_{2x}I_{3z}I_{1z}\sin(\Omega_2 t_1)\sin(\pi j_{23} t_1)\sin(\pi j_{21} t_1) \quad \text{-------(3)}$$

Only z-spin order survives following the killer gradient g3 at time point ‘e’. Thus half of the magnetizations are lost during this storage period between time point ‘d’ and ‘e’

$$+I_{2z}\cos(\Omega_2 t_1)\cos(\pi j_{23}t_1)\cos(\pi j_{21}t_1)-4I_{2z}I_{3z}I_{1z}\cos(\Omega_2 t_1)\sin(\pi j_{23}t_1)\sin(\pi j_{21}t_1)$$
$$-2I_{2z}I_{1z}\sin(\Omega_2 t_1)\cos(\pi j_{23}t_1)\sin(\pi j_{21}t_1)-2I_{2z}I_{3z}\sin(\Omega_2 t_1)\sin(\pi j_{23}t_1)\cos(\pi j_{21}t_1) \quad \text{-------(4)}$$

Now the ZS 90° pulse between time point ‘c’ and ‘d’ operates on different spins in different slices. For the slice of signal I_2, i.e. for the diagonal peak of I_2, the following terms are excited by $\pi/2(I_{2x})$

$$-I_{2y}\cos(\Omega_2 t_1)\cos(\pi j_{23}t_1)\cos(\pi j_{21}t_1)+4I_{2y}I_{3z}I_{1z}\cos(\Omega_2 t_1)\sin(\pi j_{23}t_1)\sin(\pi j_{21}t_1)$$
$$+2I_{2y}I_{1z}\sin(\Omega_2 t_1)\cos(\pi j_{23}t_1)\sin(\pi j_{21}t_1)+2I_{2y}I_{3z}\sin(\Omega_2 t_1)\sin(\pi j_{23}t_1)\cos(\pi j_{21}t_1) \quad \text{-------(5)}$$

For the slice of signal I_3, i.e. for the 2-3 cross peak, the following terms are excited by $\pi/2(I_{3x})$

$$+I_{2z}\cos(\Omega_2 t_1)\cos(\pi j_{23}t_1)\cos(\pi j_{21}t_1)+4I_{2z}I_{3y}I_{1z}\cos(\Omega_2 t_1)\sin(\pi j_{23}t_1)\sin(\pi j_{21}t_1)$$
$$-2I_{2z}I_{1z}\sin(\Omega_2 t_1)\cos(\pi j_{23}t_1)\sin(\pi j_{21}t_1)+2I_{2z}I_{3y}\sin(\Omega_2 t_1)\sin(\pi j_{23}t_1)\cos(\pi j_{21}t_1) \quad \text{-------(6)}$$

Thus only half of the magnetizations are converted into observable signals i.e. 2nd and 4th terms above in eq. (6)

For the slice of signal I_1, i.e. for the 2-1 cross peak, the following terms are excited by $\pi/2(I_{1x})$

$$+I_{2z}\cos(\Omega_2 t_1)\cos(\pi j_{23}t_1)\cos(\pi j_{21}t_1)+4I_{2z}I_{3z}I_{1y}\cos(\Omega_2 t_1)\sin(\pi j_{23}t_1)\sin(\pi j_{21}t_1)$$
$$+2I_{2z}I_{1y}\sin(\Omega_2 t_1)\cos(\pi j_{23}t_1)\sin(\pi j_{21}t_1)-2I_{2z}I_{3z}\sin(\Omega_2 t_1)\sin(\pi j_{23}t_1)\cos(\pi j_{21}t_1) \quad \text{-------} \quad (7)$$

Here also only half of the magnetizations are converted into observable signals i.e. 2nd and 4th terms above in eq. (7).

In overall, considering the total loss of magnetizations during the storage period and slice selective excitation (eq. 3 to eq. 4, and eq. 6 and 7 separately), only one fourth of the initial magnetizations contribute to the final cross peak signals. Similar calculation in G-SERF experiment will show that half of the initial magnetizations contribute to the final signal. Thus theoretically PASS-soft COSY is half as sensitive as the G-SERF experiment. In addition G-SERF experiment gains sensitivity from the refocusing nature of the experiment during t_1 evolution period in addition to the slower relaxation of the inphase signals relative to the antiphase signals that contribute to the PASS-soft COSY spectrum. Thus, the four times less sensitivity of the PASS soft COSY spectrum relative to the G-SERF spectrum in Figure 2.3 of the ms appears to agree with these calculations.

2.2(A)Beta-Butyrolactone

2.2.1 PAM Soft COSY Experiment for Beta-Butyrolactone (400 MHz)

10 mg of Beta-Butyrolactone is dissolved in 550 μL of $CDCl_3$ and data were recorded on 400 MHz spectrometer with 5mm BBO probe equipped with z-axis gradient coil only at 298 K. The duration of the ^1H 90-degree pulse was 12.95 μs.

For the duration of 41 ms, EBurp2 pulse for selective excitation and time reversal EBurp2 pulse for inversion was used (BW 102 Hz). Spectra shown in Figure 2.2A were acquired with t_1 and t_2 acquisition times of 1.46 s and 1.28 s respectively with 4 transients per t_1 increment for a total of 128 increments, TD and TD*1* were set to 4500 and 128 respectively, SW and SW*1* were set to 1600 Hz and 50 Hz. Relaxation delay was 1.2 seconds for all the experiments. Total experiment time of 29 minutes and 6 seconds. Spectra were processed with 8*K* and 512 data point in F_2 and F_1 dimensions respectively with Gaussian (GM) (SSB=2) window function and line broadening of -1.0 Hz. Frequency discrimination was achieved by using STATES-TPPI method.

The regular soft COSY spectra in Figure 2.2D and 2.2E were run with same parameters as above but using the regular soft COSY sequence of Figure 2.1B. For these two experiments the carrier of the 1st two selective pulses were put on the signal 2 while the carrier of the third selective EBurp2 pulse were shifted to the signal 3 and 1 for 2.2D and 2.2E respectively . Each experiment of the two experiments in 2.2D and 2.2E took 29 minutes and 1 seconds respectively

2.2.2 G-SERF Experiment for Beta-Butyrolactone (400MHz)

For G-SERF spectrum in Figure 2.3A, encoded E-BURP2 shape pulse had duration of 41 ms was used for the excitation that combined with the pulsed field gradient strength of 0.8 % of maximum gradient strength of 53.5 G/Cm^{-1}. The same gradient encoding was used together with a RE-BURP shaped pulse of duration 53.5 ms in the refocusing block (same bandwidth as the EBurp2 pulse). A RE-BURP pulse of duration 53.5 ms was used for the non-encoded refocusing irradiation. The offset of the non-encoded soft 180 degree pulse was set at the resonance frequency of 11^b and 20^a (labelled in a circle on the spectrum of strychnine). The coherence transfer pathway was selected by the use of sine-shaped gradient pulses of duration 1 ms and strength 50 %.

Other acquisition and processing parameters for G-SERF experiment uses the same experimental parameters as in PAM Soft COSY experiment [Figure 2.2]. Frequency discrimination was achieved by Echo-Antiecho mode. Total experiment time of 30 minutes and 30 seconds.

2.2.3 PSYCHEDELIC Experiment for Beta-Butyrolactone (400 MHz)

For PSYCHEDELIC spectrum in Figure 2.3B, Chirp pulses for PSYCHE element had bandwidth of 10 KHz and duration of 15 ms each, with a maximum RF amplitude of 36Hz and flip angle of 11.5°. The pulsed field gradient strength of 1.5 % of maximum gradient strength of 53.5 G/Cm^{-1} used during PSYCHE element. The coherence transfer pathway was selected by the use of SMSQ shaped gradient pulses of duration 1 ms and strength 20 % of maximum gradient strength. For the BIP pulses, BIP 720, 50, 20.1 pulses of 300 us were used, and the power level calculated based on the calibrated hard 90° pulse. For the encoded selective shaped pulse, RE-BURP2 180° pulse was used, with duration of 35 ms.

For the Beta-Butyrolactone, spectra shown in Figure 2.3B, SW were set to 1600 Hz (4 ppm) in the direct (F_3) dimension, 32 Hz in the indirect scalar coupling dimension (F_1) dimension, and 40 Hz in the interferogram pure shift dimension (F_2). The number of both t_1 and t_2 increments (number of chunks) were set to 32 for the full PSYCHEDELIC experiment. 4 transients for all the increments. t_3, t_2, and t_1 acquisition times were 800ms, 500 ms, and 400 ms respectively. Spectra were processed with $32K$ and 128 data point in F_2 and F_1

dimensions respectively with Gaussian (GM) (SSB=2) window function and line broadening of -1.0 Hz. Frequency discrimination was achieved by using Echo-Antiecho mode in F_2 and QF mode in F_1 dimension respectively. Total experiment time of 4 hour 7 minutes.

2.3 (B) Strychnine

2.3.1 PAM Soft COSY Experiment for Strychnine (400 MHz)

30 mg of Strychnine is dissolved in 550 μL of CDCl3 and data were recorded on 400 MHz spectrometer with 5mm BBO probe equipped with z-axis gradient coil only at 298 K. The duration of the ^{1}H 90-degree pulse was 12.95 μs.

For the duration of 35 ms, EBurp2 pulse for selective excitation and time reversal EBurp2 pulse for inversion was used. Spectra shown in Figure 2.4A were acquired with t_1 and t_2 acquisition times of 1.46 s and 1.40 s respectively with 4 transients per t_1 increment for a total of 176 increments, TD and TD*1* were set to 4500 and 176 respectively, SW and SW*1* were set to 1600 Hz and 60 Hz. Relaxation delay was 1.2 seconds for all the experiments. Total experiment time of 41 minutes and 30 seconds. Spectra were processed with 8*K* and 512 data points in F_2 and F_1 dimensions respectively with Gaussian (GM) (SSB=2) window function and line broadening of -1.0 Hz. Frequency discrimination was achieved by using STATES-TPPI method.

2.3.2 G-SERF Experiment for Strychnine (400MHz)

For G-SERF spectrum in Figure 2.4F, encoded E-BURP2 shape pulse had duration of 35 ms was used for the excitation that combined with the pulsed field gradient strength of 0.8 % of maximum gradient strength of 53.5 G/Cm^{-1}. The same gradient encoding was used together with a RE-BURP shaped pulse of duration 35 ms in the refocusing block. A RE-BURP pulse of duration 35 ms was used for the non-encoded refocusing irradiation. The offset of the non-encoded soft 180 degree pulse was set at the resonance frequency of 11^b and 20^a (labelled in a circle on the spectrum of strychnine). The coherence transfer pathway was selected by the use of sine-shaped gradient pulses of duration 1 ms and strength 50 %.

Spectra acquired and processed for G-SERF experiment in 2.4F uses the same experimental parameters as in PAM Soft COSY. Frequency discrimination was achieved by Echo-Antiecho mode. Total experiment time of 43 minutes and 30 seconds.

2.3.3 PSYCHEDELIC Experiment for Strychnine (400 MHz)

For PSYCHEDELIC spectrum in Figure 2.4G, Chirp pulses of used in PSYCHE element had bandwidth of 10 KHz and duration of 15 ms each, with a maximum RF amplitude of 46 Hz and flip angle of 11.5°. The pulsed field gradient strength of 1.5 % of maximum gradient strength of 53.5 G/Cm^{-1} used during PSYCHE element. The coherence transfer pathway was selected by the use of SMSQ shaped gradient pulses of duration 1 ms and strength 20 % of maximum gradient strength. For the BIP pulses, BIP 720, 50, 20.1 pulses of 300 us were used, and the power level calculated based on the calibrated hard 90° pulse. For the encoded selective shaped pulse, RE-BURP2 180° pulse was used, with duration of 35 ms.

For the strychnine, spectra shown in Figure 2.4G were, SW were set to 1600 Hz (4 ppm) in the direct (F_3) dimension, 50 Hz ($=1600/2^5$) in the indirect scalar coupling dimension (F_1) dimension, and 40 Hz in the interferogram pure shift dimension (F_2). The number of both t_1 and t_2 increments (number of chunks) were set to 32 for the full PSYCHEDELIC experiment. 4 transients for all the increments. t_3, t_2, and t_1 acquisition times were 1.6 sec, 320 ms, and 400 ms respectively. Spectra were processed with 32*K* and 128 data point in F_2 and F_1 dimensions respectively with Gaussian (GM) (SSB=2) window function and line broadening of -1.0 Hz. Frequency discrimination was achieved by using Echo-antiecho method in F_2 and QF mode in F_1. Total experiment time of 4 hour 7 minutes.

2.4 (C)Hesperidin

2.4.1 PAM Soft COSY for Hesperidin

10 mg of Hesperidin is dissolved in 550 μL of CDCl3 and data were recorded on 800 MHz spectrometer equipped with CPTCI cryo probe and z-axis gradient only at 300 K. Selective

excitation of spins was achieved by using EBurp2 pulse of 50 ms duration, whereas the 50 ms long time reversal pulse for the inversion of spins is used.

Spectra in Figure 2.5A were acquired with t_1 and t_2 acquisition times of 500 ms respectively with 4 transients per t_1 increment for a total of 100 increments, TD and TD*1* were set to 9998 and 100 respectively, SW and SW*1* were set to 10000 Hz and 100 Hz. Total experiment time of 16 minutes. Spectra were processed with 16*K* and 256 data point in F_2 and F_1 dimensions respectively with shifted sine bell (SSB=2) window function and line broadening of 0.3Hz. Frequency discrimination was achieved by using STATES-TPPI method.

3. Results and discussions

The pulse scheme for acquiring PASS-soft-COSY spectrum is displayed in Fig. 2.1A. It involves application of three selective pulses. Three signals labelled as I_1, I_2, and I_3 as shown in Fig. 2.1C are considered with resonance frequencies m1, m2, and m3 respectively. The 1st selective pulse between time points 'a' and 'b' is applied on signal I_1 with resonance frequency m1. During t_1 this signal evolves under its own chemical shift and scalar coupling interactions to signals I_2, and I_3. The second selective pulse between time points 'c' and 'd' is also applied on the signal I_1. The dephasing gradient g_3 between time points'd' and 'e' removes all the transverse components of magnetization.

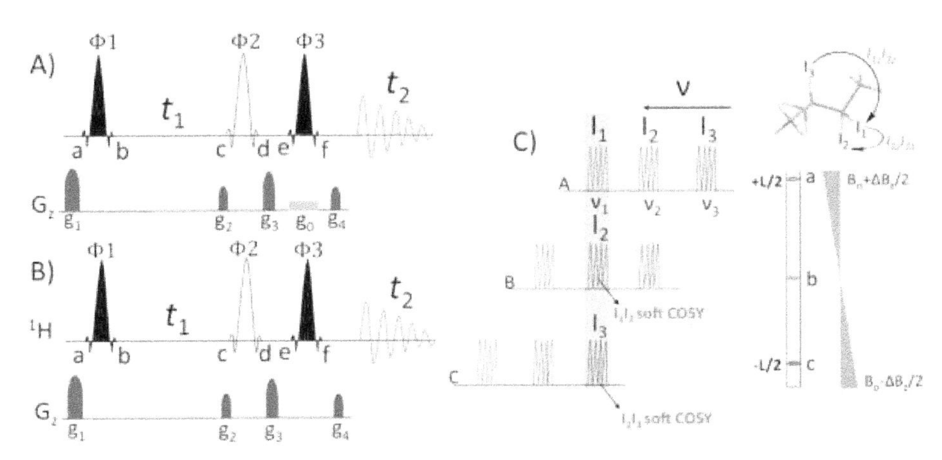

Fig. 2.1. (A) Pulse sequence for 2D PASS-soft-COSY. Three shaped pulses are applied between time points 'a' and 'b' (EBurp2), 'c' and 'd' (Time Reversed EBurp2), and 'e' and 'f' (EBURP2). g_1, g_2, g_3, and g_4 are SINE shaped gradient, while g_0 is a rectangular gradient. The third selective

pulse between time points 'e' and 'f' together with the simultaneously applied gradient g_0 constitutes the ZS 90° pulse sequence element. g_1 and g_3 are crusher gradient. g_2 and g_4 are coherence selection gradient. The various phases for the used pulses are: Φ_1= x -x, Φ_2= x, x, -x, -x, Φ_3= x, x, x, x, and receiver phase Φ_{rec}= x, -x,-x, x. Gradient strength were G_1=30 %, G_2=18 %, G_3=18 %, G_4=43 %, G_0=1.5%. All gradients are SINE bell shifted with 1 ms of duration, whereas the slice selective gradient applied for 50 us duration, and rectangular in shape. (B) Pulse sequence for regular soft COSY. The pulses and gradients displayed are kept same as in PASS-soft-COSY. (C) Schematic description of the spatial encoding involved in PASS-soft-COSY. Three coupled signals labelled as I_1, I_2, and I_3 have frequencies m1, m2, and m3 respectively. Selective excitation, t_1 encoding, and storage of the signal I_1 only is carried out by the 1st and 2nd semi-selective pulses from the time point 'a' to 'd'. The dephasing gradient g_3 removes all the transverse components of magnetization. Only I_{1Z}, $I_{1Z}I_{2Z}$, and $I_{1Z}I_{3Z}$ spin orders survive at time point 'e'. During the ZS 90° pulse sequence element the resonance frequencies of the signals I_1, I_2, and I_3 undergo a location dependent frequency shift as shown in the three spectra A, B and C in (C). The coherence transfers I_1 to I_2, and I_1 to I_3 via the terms $I_{1Z}I_{2Z}$ and $I_{1Z}I_{3Z}$ are now executed in a slice selective manner. The two terms $I_{1Z}I_{2Y}$ and $I_{1Z}I_{3Y}$ leading to two separate soft COSY spectra are now created in separate slices B and C at the start of data acquisition. In each of these two separate slices, the spin state of the other passive spins are not disturbed. Thus the resulting spectrum will contain the two soft COSY spectrum and a diagonal peak of I_1.

The magnetizations that are left at the time point 'e' are I_{1Z}, $I_{1Z}I_{2Z}$, and $I_{1Z}I_{3Z}$ i.e. only the z-spin orders survive. In a regular soft COSY pulse sequence shown in Fig. 2.1B, the third 90° pulse is selectively applied either on signal I_2 or I_3, so that reduced multiplet pattern is generated which leads to a simplified spectrum. Because the third selective pulse needs to be applied on only one of the two coupled signals ($I_{1Z}I_{2Z}$ or $I_{1Z}I_{3Z}$), the regular soft COSY approach will demand two separate 2D experiments to get simplified spectrum. However, in the present scheme a ZS 90° pulse sequence element is applied between time points 'e' and 'f' (Fig. 2.1A). This serves the purpose of all the separate soft COSY experiments related to signal 1 in a single experiment. Because the ZS pulse sequence element applies selective pulse in the presence of a weak magnetic field gradient (g_0), the resonance frequencies of the signals I_1, I_2, and I_3 undergo a location dependent frequency shift as shown in the three spectra A, B and C in Fig. 2.1C. The coherence transfer from signal 1 to a second spin (i.e. I_2 or I_3) via the terms $I_{1Z}I_{2Z}$ and $I_{1Z}I_{3Z}$ (between time points 'e' and 'f') are now executed in a slice selective manner. In other words, the ZS 90° pulse sequence element will create the two terms $I_{1Z}I_{2Y}$ and $I_{1Z}I_{3Y}$ in separate slices B and C at the start of data acquisition. In each of these two separate slices, the spin states of the other passive spins are not disturbed preserving the spin state selective coherence transfer of the soft COSY experiment. The two signals from the two slices B and C corresponding to the terms $I_{1Z}I_{2Y}$ and $I_{1Z}I_{3Y}$ will contribute to the final spectrum in addition to the I_{1Y} term (via I_{1Z}) from the slice A. Thus the resulting PASS- soft COSY spectrum will contain two soft COSY spectra (I_1-I_2 and I_1-I_3 selective correlations) in addition to the diagonal peak of I_1. In Fig. 2.1A, g_1 and g_3 are crusher gradient, g_2 and g_4 are coherence selection gradient.

3.1. Results and Discussions for strychnine β-butyrolactone

An experimental PASS-soft-COSY spectrum is shown in Fig. 2.2A for the proton 2 in β-butyrolactone which is scalar coupled to the signals 1 and 3 as shown in the molecular structure (ignoring long range scalar coupling $^4J_{HH}$ to signal 4). The 1st two selective pulses are applied on signal 2 so that chemical shift and scalar couplings of signal 2 is encoded in the F_1 dimension. Subsequently, the ZS mixing pulse is applied so that all the signals 1–4 are excited in separate slices. However, only the anti-phase spin orders i.e. $I_{2Z}I_{1Z}$ and $I_{2Z}I_{3Z}$ (from separate slices) which are present between time points 'd' and 'e' and follow the precise coherence transfer pathway (of COSY) generate the 2–1 and 2–3 soft COSY spectra shown as shaded box labelled 'q' and 'p' respectively in Fig. 2.2A. The 2–3 and 2–1 soft COSY spectra from Fig. 2.2A are shown expanded in Fig. 2.2B and C respectively. Separate soft COSY spectrum of signals 2–3 and 2–1 were also recorded for comparison (using the regular soft COSY pulse sequence of Fig. 2.1B) and shown in Fig. 2.2D and E respectively. Identical results can be noted from the comparison of Fig. 2.2B vs. D, and C vs. E. Thus in a single experiment two soft COSY spectra could be generated using spatially selective mixing saving precious spectrometer time by 50%.

The analysis of the 2–3 soft COSY (Fig. 2.2B) spectrum provides the active coupling a = $^2J_{H2H3}$ (=16.4 Hz) along both dimensions forming a regular square pattern shown as rectangular box. The passive coupling b = $^3J_{H1H2}$(=4.2 Hz) leads to displacement of the square boxes along F_1 and thus value of this passive coupling can be measured along F_1. Similarly the other passive coupling c = $^3J_{H1H3}$(=5.8 Hz) can be measured from the displacement of the boxes along F_2. The line-shape is antiphase with respect to the active coupling and in phase with respect to the passive coupling. Thus, the simplified 2D spectral pattern of the 2–3 part of the PASS-soft-COSY spectrum provides a simpler analysis for extraction of the scalar couplings. Then we focus on the analysis of the complex multiplet of the signal 1 using the 1–2 part of the PASS-soft-COSY spectrum.

The analysis of the 2–1 soft COSY spectrum (Fig. 2.2C) provides the active coupling b = $^{3}J_{H1H2}$ (=4.2 Hz) along both dimensions forming a regular square pattern shown as dotted box on the right bottom corner. The passive couplings for this 2–1 cross peaks are c = $^{3}J_{H1H3}$, e = $^{3}J_{H1H4}$, a = $^{3}J_{H2H3}$ and d = $^{4}J_{H2H4}$. The passive coupling a = $^{2}J_{H2H3}$ (=16.4 Hz) leads to displacement of the square boxes along F_1 and thus value of this passive coupling can be measured along F_1. Similarly the other passive coupling along F_1, d = $^{4}J_{H2H4}$ is a long range coupling and leads to very small displacement and could not be measured due to line-shape distortions. The other two passive couplings along F_2 i.e. c = $^{3}J_{H1H3}$(=5.8 Hz) and e = $^{3}J_{H1H4}$ (=5.8 Hz) can be measured from the displacement as shown in Fig. 2.2C. Thus, the simplified 2D spectral pattern of the 2–1 part of the PASS-soft COSY spectrum provides a simpler deconvolutions of the complex multiplet pattern of signal 1.

G-SERF and PSYCHEDELIC experiments were also recorded and displayed in Fig. 2.3A and B respectively. G-SERF yields the active couplings a = $^{2}J_{H2H3}$ (=16.3 Hz), and b = $^{3}J_{H1H2}$(=4.3 Hz) of the signal H2 with signals H1 and H3 respectively. While, PSYCHEDELIC yields the active couplings a = $^{2}J_{H2H3}$ (=16.3 Hz) and b = $^{3}J_{H1H2}$(=4.4 Hz) respectively. No passive coupling information of H1 and H3 with other spins is provided by these spectra. In contrast the PASS-soft-COSY spectrum in Fig. 2.2A provides two passive coupling information viz. c = $^{3}J_{H1H3}$(=5.8 Hz) e = $^{3}J_{H1H4}$ (=5.8 Hz) in addition to the above two active couplings.

We also made a comparison of the sensitivity of the three experiments - PASS-soft-COSY, G-SERF, and regular Soft COSY. This comparison is displayed in Fig. 2.3C. The intensity of the peaks in G-SERF experiment was six times less than regular soft COSY experiment. Further, the intensity of the peaks in PASS-soft COSY experiment was twenty seven times less than regular soft COSY experiment. Thus sensitivity of the PASS-soft-COSY spectrum is found to be less than G-SERF by a factor close to four. This is because in PASS-soft-COSY magnetization from a given spin is converted into four terms for each additional coupling whereas in G-SERF it is converted into two terms. While only one of the four terms contributes to the final observable spectrum in PASS-soft COSY, one of the two terms contributes in G-SERF experiment. This calculation for sensitivity is detailed in the experimental

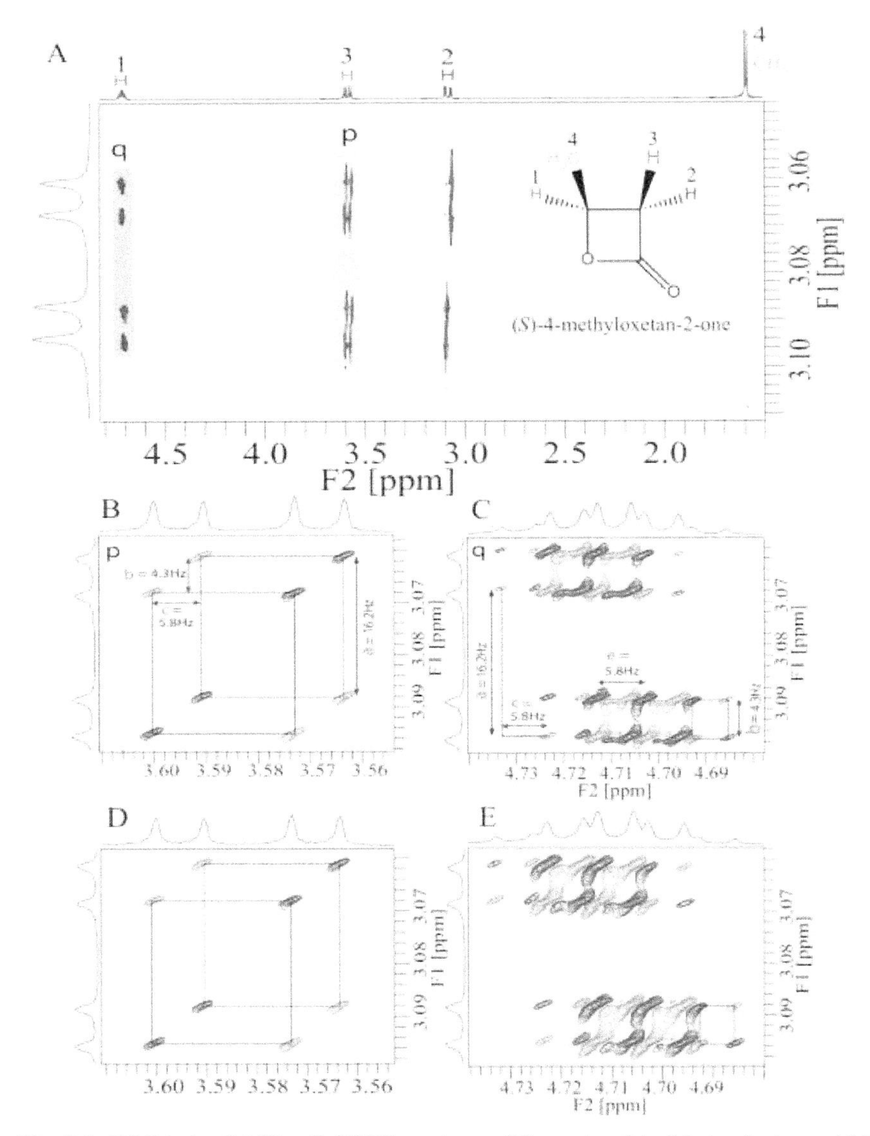

Fig. 2.2. (A) Display PASS-soft-COSY spectrum of the proton 2 in β-butyrolactone which is scalar coupled to the signals 1 and 3. Spectrum from only signal 2 is recorded in the F_1 dimension. The two soft COSY spectra 2–1 and 2–3 shown as shaded box labelled 'q' and 'p' respectively are acquired simultaneously. The 2–3 (p) and 2–1 (q) soft COSY spectra from (A)

are further shown expanded in (B) and (C) respectively. (D) and (E) are the separate soft COSY spectrum of signals 2–3 and 2–1 recorded for comparison using the regular soft COSY pulse sequence of (1B). Identical results can be noted from the comparison. Thus in a single experiment two soft COSY spectra could be generated using spatially selective mixing saving precious spectrometer time by 50%. The analysis of the 2–3 soft COSY spectrum (B) provides the active coupling a = $^{2}J_{H2H3}$ (=16.2 Hz) along both dimensions forming a regular square pattern. The passive coupling b = $^{3}J_{H1H2}$(=4.3 Hz) leads to displacement of the square boxes along F_1. Similarly the other passive coupling c = $^{3}J_{H1H3}$(=5.8 Hz) can be measured from the displacement of the boxes along F_2. Thus, the 2–3 part of the parallel soft COSY spectrum provides a simpler analysis for extraction of the scalar couplings. The analysis of the 2–1 part (C) provides the active coupling b = $^{3}J_{H1H2}$ (=4.3 Hz) and the passive couplings c = $^{3}J_{H1H3}$(=5.8Hz), e = $^{3}J_{H1H4}$ (=5.8Hz), a = $^{2}J_{H2H3}$ (=16.2 Hz) and d = $^{4}J_{H2H4}$ (which is a long range coupling and leads to very small displacement, line shape distortions prevented its measurement). Thus the simplified spectral pattern of the 1–2 part of the parallel soft COSY spectrum provides a simpler de-convolutions of the complex multiplet pattern of signal 1.

section product operators. In addition G-SERF experiment gains sensitivity from the refocusing nature of the experiment during t_1 evolution period in addition to the slower relaxation of the inphase signals relative to the antiphase signals that contribute to the PASS-soft COSY spectrum. All these factors collectively lead to four times less sensitivity for the PASS-soft COSY experiment. There are various ways to enhance the sensitivity of the ZS experiments. One option is to use short selective pulse to increase the size of the slices. Other options are equidistant and non-equidistant n-fold modulations of the selective pulse used in the ZS element[8,10], or generating frequencies for simultaneous multi-slice utilization tailored to a specific sample[9]. Introduction of frequency shifted selective pulses for the ZS pulse sequence element has also been reported to enhance sensitivity per unit of measurement time[30]. Despite low sensitivity, the PASS-soft COSY can reduce experimental time drastically. It should be noted that the PASS-soft COSY spectrum in Fig. 2.2A was recorded in 29 min while the two separate soft COSY spectra of Fig. 2.2D and E took a total of 58 min spectrometer time. Thus a 50% decrease in experimental time can be noted. Thus where the S/N is outstanding or adequate, like the present case, PASS-soft COSY approach will save precious spectrometer time. Because 2D experiments regardless of sensitivity, require large number of scans to sample the increment based t_1 time domain data[18,19]. However, for low concentration samples, such reduction in experimental time with a low sensitive technique will not be beneficial. More signal averaging to get a reasonable S/N will outweigh the advantage of parallel acquisition.

Another point worth highlighting is the antiphase line shape of the peaks w. r. to the active coupling. If the active couplings are very small in magnitude, self-cancellation of the antiphase multiplet components can obscure the lineshape. This will generally happen in spin systems with long range couplings, and only when transfer of magnetizations to remotely coupled signals will be attempted. In addition, another complication can arise from the self-cancellation of the antiphase multiplet

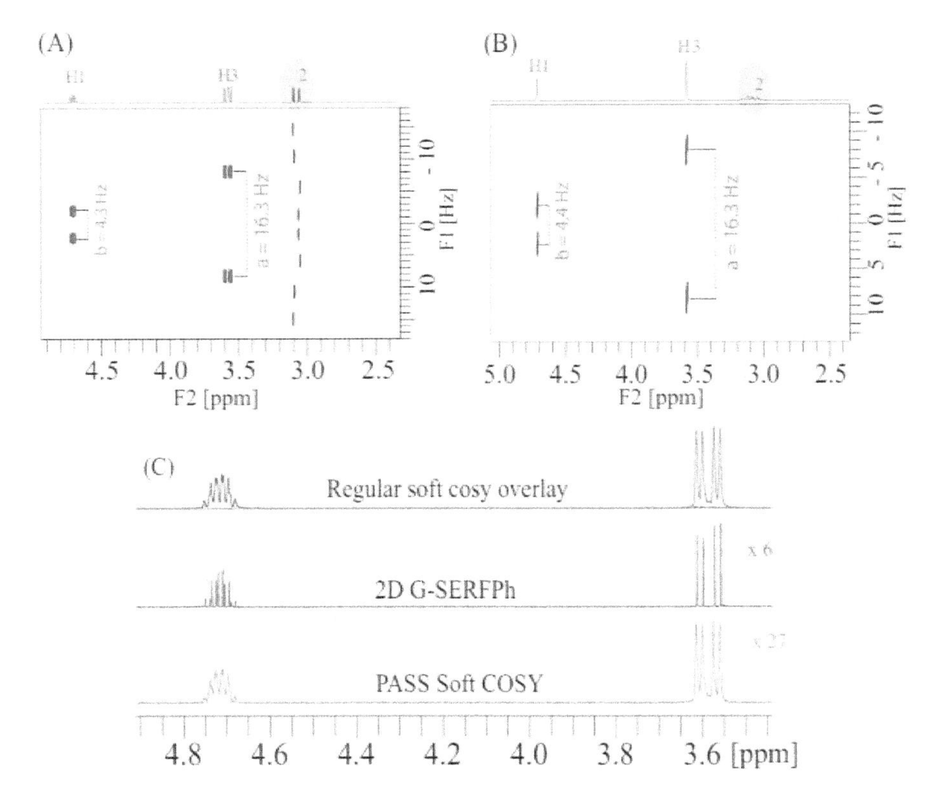

Fig. 2.3. (A) and (B) display the G-SERF and PSYCHEDELIC spectra respectively. The G-SERF spectrum yields the active couplings a = $^2J_{H2H3}$ (=16.3 Hz), and b = $^3J_{H1H2}$(=4.3 Hz). The PSYCHEDELIC spectrum also yields a = $^2J_{H2H3}$ (=16.3 Hz), and b = $^3J_{H1H2}$(=4.4 Hz) respectively, with negligible difference in the measurement of couplings. No passive coupling information of H1 and H3 with other spins is provided by these spectra. (C) display the comparison of the sensitivity of the three experiments PASS-soft-COSY, G-SERF, and regular Soft-COSY from

bottom to top respectively. The intensity of the peaks in G-SERF and PASS-soft COSY experiment were found to be six times and twenty seven times less than those in the regular soft COSY experiment respectively. Thus sensitivity of the PASS-soft-COSY experiment is found to be less than the G-SERF by a factor close to four. The reason for this difference in sensitivity is detailed in the experimental section product operators.

Table 2.1 Coupling constant measured by PAM Soft COSY, G-SERF, PSYCHEDELIC Experiment for Beta-Butyrolactone

Types of Proton	PAM Soft COSY $^{n}J_{HH}$(Hz)	2D G-SERF $^{n}J_{HH}$(Hz)	PSYCHEDELIC $^{n}J_{HH}$(Hz)
$^{2}J_{H2\text{-}H3}$ = a	16.4Hz	16.3Hz	16.3Hz
$^{3}J_{H1\text{-}H2}$ = b	4.2Hz	4.3Hz	4.4Hz
$^{3}J_{H1\text{-}H3}$ = c	5.8Hz	NA	NA
$^{3}J_{H1\text{-}H4}$ = e	5.8Hz	NA	NA

3.2 Results and Discussions for strychnine

In another application of the PASS-soft-COSY scheme, three soft COSY spectra were recorded simultaneously on strychnine. Selective excitation and storage of the signals H_{20a} and H_{11b} (as they are very close in frequency) ensured only their chemical shifts and J-couplings to be encoded during F_1 as displayed in Figure 2.4B. Such simultaneous chemical shift encoding of different signals along the indirect dimension is not feasible in SERF series of experiments due to refocusing of chemical shift interactions. Subsequent application of the ZS mixing element produced three soft COSY spectra marked as L, M, and N in Figure 2.4B. The soft COSY spectra L, M, and N correspond to H_{11b}-H_{11a}, H_{20a}-H_{20b}, and H_{11b}-H_{12}

selective proton-proton correlations respectively and are shown in expanded form in Figures 2.4C, 2.4D, and 2.4E respectively. Analysis of the soft COSY spectrum L (H_{11b}-H_{11a}) as shown in Figure 2.4D yields three couplings a(17.4 Hz, $^{2}J_{H11b\text{-}H11a}$, active coupling), b(3.2 Hz, $^{3}J_{H11a\text{-}H12}$, passive coupling), c(8.4 Hz, $^{3}J_{H11b\text{-}H12}$, passive coupling). Analysis of the soft COSY spectrum N (H_{11b}-H_{12}) as shown in Figure 2.4E yields one more passive coupling d(3.2 Hz, $^{3}J_{H12\text{-}H13}$) in addition to the couplings a, b, and c as already measured in spectrum L also.

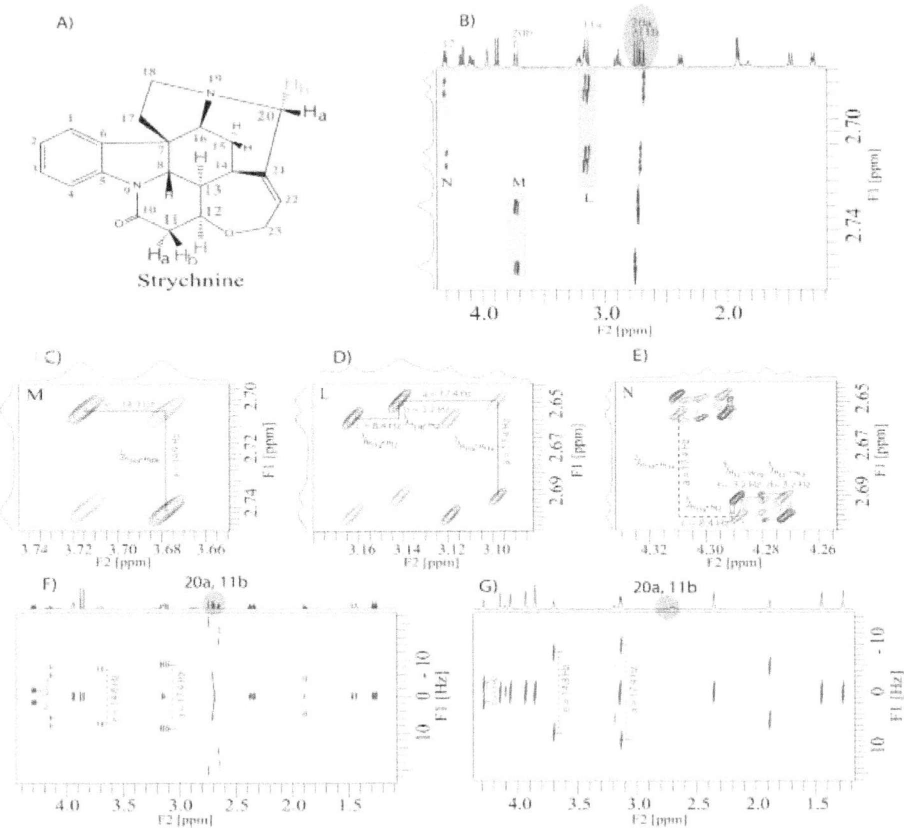

Fig. 2.4. Figure (A) display PAM-soft-COSY spectrum of the protons H_{20a} and H_{11b} in strychnine molecule. Only the chemical shifts and the *J*-couplings of these two signals are recorded along F_1 by applying all the selective pulses on these two signals as they are very close in frequency. Three soft COSY spectrum marked as L, M, and N are obtained simultaneously. The soft COSY spectra L and N correspond to H_{11b}-H_{11a}, H_{11b}-H_{12} selective proton-proton correlations and shown expanded in Figures 2.4D and 2.4E respectively. The other selective H_{20a}-H_{20b} correlation generates the soft COSY spectrum M and shown expanded in Figure 2.4C.

Analysis of the soft COSY spectrum L as shown in Figure 2.4D yields three couplings a(17.3 Hz, $^{2}J_{H11b\text{-}H11a}$, active coupling), b(3.3 Hz, $^{3}J_{H11a\text{-}H12}$, passive coupling), c(8.5 Hz, $^{3}J_{H11b\text{-}H12}$, passive coupling). Analysis of the soft COSY spectrum N as shown in Figure 2.4E yields one more passive coupling d(3.3 Hz, $^{3}J_{H12\text{-}H13}$). As the active and passive couplings b and d are similar in magnitude they tend to cancel out the central lines of the multiplet leading to slightly distorted peaks. The soft COSY spectrum M between signals H_{20a} and H_{20b} in Figure 2.4C is from another part of the strychnine molecule which does not have any couplings to protons involved in L and N part. The active coupling between these two geminal protons form a square pattern as displayed in 2.4C. There exists five long range passive couplings ($^{4}J_{HH}$) for each of the signals H_{20a} and H_{20b} which leads to the tilted pattern adding up to the line-width combined with the field inhomogeneity.

As the active and passive couplings b and d are similar in magnitude they tend to cancel out in the middle leading to distorted peaks. This is discussed in the main manuscript as well. The soft COSY spectrum M between signals H_{20a} and H_{20b} in Figure 2.4C is from an entirely different part of the strychnine molecule. The active coupling between these two geminal protons form a square pattern as expected. Five long range passive couplings ($^{4}J_{HH}$) to H_{20a} and H_{20b} adds up to the line-width combined with field inhomogeneity and generate the tilted pattern for each of the peaks in Figure 2.3B.

Table 2.2 Coupling constant measured by PAM Soft COSY, G-SERF, PSYCHEDELIC Experiment for strychnine

Types of Proton	PAM Soft COSY $^{n}J_{HH}$(Hz)	2D G-SERF $^{n}J_{HH}$(Hz)	PSYCHEDELIC $^{n}J_{HH}$(Hz)
$^{2}J_{H11a\text{-}H11b}$=a	17.4	17.4	17.4
$^{2}J_{H20a\text{-}H20b}$=e	14.9	14.8	14.8
$^{3}J_{H11b\text{-}H12}$=b	3.2	3.4	3.4
$^{3}J_{H12\text{-}H13}$=d	3.2	NA	NA
$^{3}J_{H11a\text{-}H12}$=c	8.4	NA	NA

3.3. Results and Discussions for hesperidin molecule

The third application of the PASS-soft-COSY scheme is demonstrated on hesperidine molecule, where three soft COSY spectra were recorded simultaneously. Selective excitation and storage of the signals H_{3a}, $H_{2''}$, and $H_{3''}$ (as they are unresolved) ensured only their chemical shifts and *J*-couplings to be recorded during F_1 as displayed in Figure 2.5A. The horizontal black line on top shows the correlation between $H_{2''}$-$H_{1''}$. The other horizontal line in red display the selective correlation between H_{3a}-H_{3b}, and H_{3a}-H_2 protons. Thus three soft COSY spectra G, H, and I corresponding to H_{3a}-H_{3b}, H_{3a}-H_2, and $H_{2''}$-$H_{1''}$ selective proton-proton correlations were acquired parallelly, and are shown in expanded form in Figures 2.5B, 2.5C, and 2.5D respectively. Analysis of the soft COSY spectrum G (H_{3a}-H_{3b}) as shown in Figure 2.5B yields three couplings a (17.1 Hz, $^2J_{H3a\text{-}H3b}$, active coupling), b (12.1 Hz, $^3J_{H3a\text{-}H2}$, passive coupling), c (3.3 Hz, $^3J_{H3b\text{-}H2}$, passive coupling). It is to be noted that the signals corresponding to H_{3a} and H_{3b} protons are splitted into two sets of signals with slightly different chemical shifts due to relative configurational isomers (flavones such as hesperidine naturally exist in two different isomeric configurations at C-2 position)[27-28]. This scales up the crowding of the signals and hence complexity. However, the soft COSY pattern deconvolutes the signals into two well resolved sets and aids easier assignment.

Similarly analysis of the soft COSY spectrum H (H_{3a}-H_2) as shown in Figure 2.5C also yields the same couplings as in G although the designation of active and passive couplings are different i.e. a (17.1 Hz, $^2J_{H3a\text{-}H3b}$, passive coupling), b (12.1 Hz, $^3J_{H3a\text{-}H2}$, active coupling), c (3.3 Hz, $^3J_{H3b\text{-}H2}$, passive coupling). The long range couplings $^4J_{HH}$ from H_2 to $H_{6'}$ and $H_{2'}$ are unresolved due to line broadening. Two sets of signals are again visible and simplified spectral pattern is evident. Analysis of the soft COSY spectrum I ($H_{2''}$-$H_{1''}$) is shown in Figure 2.5D yields the couplings d (7.4 Hz, $^3J_{H2''\text{-}H1''}$, active coupling). The other long range passive couplings of $H_{1''}$ could not be measured due to line-width.

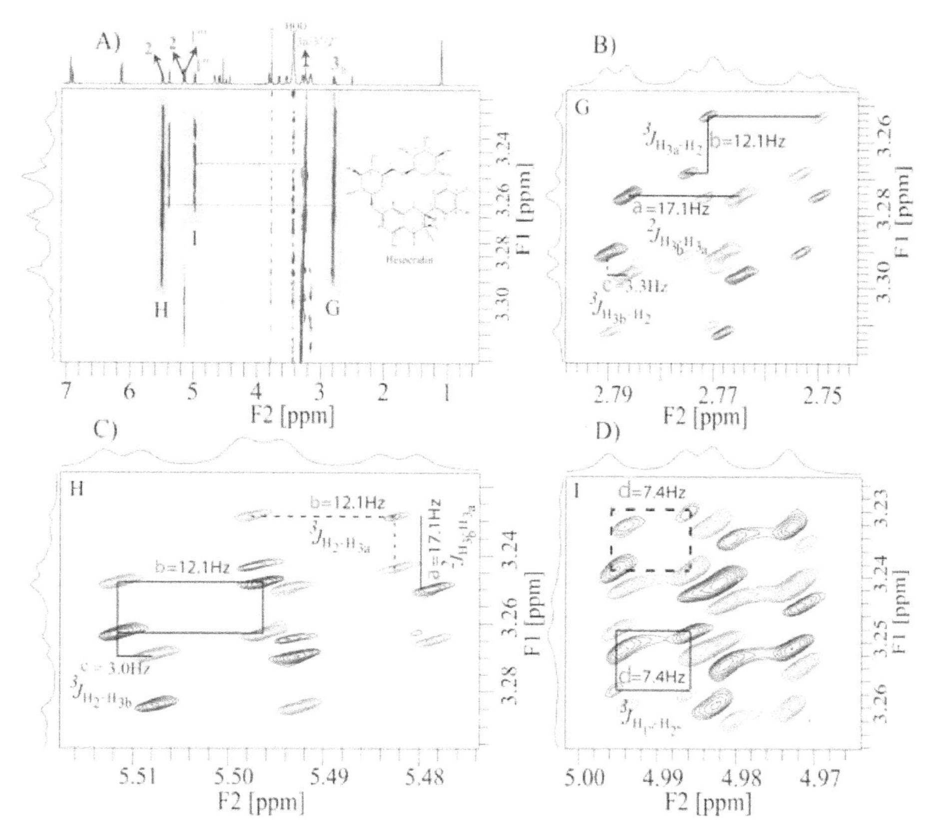

Fig. 2.5 . Figure (A) display PASS-soft-COSY spectra corresponding to the H_{3a}-H_{3b}, H_{3a}-H_2, and $H_{2''}$-$H_{1''}$ correlations in hesperidine molecule. Selective excitation and storage of the signals H_{3a}, $H_{2''}$, and $H_{3''}$ (as they are unresolved) ensured only their spectrum to be recorded along F_1. Thus three soft COSY spectra labelled as G, H, and I corresponding to H_{3a}-H_{3b}, H_{3a}-H_2, and $H_{2''}$-$H_{1''}$ selective proton-proton correlations could be recorded simultaneously, and are shown in expanded form in Figures 2.5B, 2.5C, and 2.5D respectively. Analysis of the soft COSY spectrum G (H_{3a}-H_{3b}) as shown in Figure 2.5B yields three couplings a (17.1 Hz, $^2J_{H3a\text{-}H3b}$, active coupling), b (12.1 Hz, $^3J_{H3a\text{-}H2}$, passive coupling), c (3.3 Hz, $^3J_{H3b\text{-}H2}$, passive coupling). Two sets of signals for each of H_{3a} and H_{3b} are observed with slightly different chemical shifts due to relative configurational isomers (at C-2 position). This leads to more crowded signals. However, the simplified soft COSY pattern aids easier assignment and extraction of the couplings. The soft COSY spectrum H (H_{3a}-H_2) as shown in Figure 2.5C also yields the same couplings as in G. The long range couplings $^4J_{HH}$ from H_2 to $H_{6'}$ and $H_{2'}$ are unresolved due to line broadening. Two sets of signals are again visible and simplified spectral pattern is evident. Analysis of the soft COSY spectrum I ($H_{2''}$-$H_{1''}$) as shown in Figure 2.5D yields the coupling d(7.4 Hz, $^2J_{H2''\text{-}H1''}$, active coupling), The other long range passive couplings of H1'' could not be measured due to line-broadening.

4. Conclusion

Thus, we have demonstrated a novel pulse technique for simultaneous acquisition of several slice selective soft COSY spectra in a single experiment. A ZS 90° pulse sequence element used during a mixing period enables spatially encoded coherence transfer which is crucial part of this new scheme. Scalar coupling between the excited signal and a coupled neighbour can be measured directly on the signal of the coupled neighbour as active coupling forming a square pattern. In addition passive couplings of the excited signal or of the coupled neighbour with any other signals can be mea- sured from the displacements along F_1 and F_2. The long range passive couplings being very small in magnitude and comparable to linnewidths lead to line broadening.

Although, the G-SERF experiment also allows parallel acquisition of several SERF spectra, however, the fundamental difference with PASS-soft-COSY is that, it is not a coherence transfer scheme. The drawback of the PASS-soft COSY experiment is that it is found to be nearly four times less sensitive than the G-SERF experiment, and self-cancellation of antiphase components in certain case can distort the peaks. However, it underscores other advantages such as feasibility of encoding multiple chemical shift in the t_1 period as it allows chemical shift evolution unlike G-SERF, and ability to measure passive couplings. Since the ZS pulse sequence element has sensitivity which is only 2-3 percent of the regular proton NMR, therefore, these experiments will be suitable when sample concentration is not too low. In addition, methods to improve sensitivity of ZS technique have been reported. Considering today's high field magnets, cryoprobe technology, and spin hyperpolarization techniques, such experiments are becoming feasible. The method is also applicable to various other multi-dimensional selective correlation techniques.

5. Acknowledgments

We thank SERB (DST) as this work was supported by extra mural research fund provided by Science and Engineering Research Board under Department of Science & Technology, Govt. of India (Grant No. SERB/F/6435/2015-16). Upendra Singh thanks University Grants Commission (UGC) for Research fellowship. Ajay Verma thanks Council of Scientific and

Industrial Research (CSIR), India, for research fellowship. We thank the Director, CBMR for research facilities.

6. Future Work

The homonuclear scalar coupling constant values play an important role in the structural identification of molecules and identification of metabolites. But, the measurement of homonuclear scalar coupling constant is very difficult due to overlapping of multiplet signals in the natural products, and bio-fluids such as serum, urine, saliva etc. These bio-fluids contain number of primary and secondar metabolites which play an important role in the living organism. We will try to utilize this method in the measurement of scalar coupling constant values to provide crucial information to identify primary and secondary metabolites in bio-fluids and plant extracts by using the data of Human Metabolome Database (HMDB), and Biological Magnetic Resonance Bank (BMRB).

References

[1] K. Zangger, H. Sterk. Homonuclear Broadband-Decoupled NMR Spectra, *J. Magn. Reson.* 489, (1997), 486–489.

[2] K. Zangger, Pure shift NMR, Prog. NMR Spectrosc. 86–87 (2015) 1–20.

[3] N. H. Meyer, K. Zangger. Simplifying Proton NMR Spectra by Instant Homonuclear Broadband Decoupling, Angew. Chem. Int. Ed. 52, (2013), 7143-7146.

[4] P. Sakhaii, B. Haase, W. Bermel, R. Kerssebaum, G. E. Wagner, K. Zangger. Broadband homodecoupled NMR spectroscopy with enhanced sensitivity, J. Magn. Reson. 233, (2013), 92–95.

[5] N. Lokesh, N. Suryaprakash, Sensitivity enhancement in slice-selective NMR experiments through polarization sharing, Chem. Commun. 50, (2014), 8550–8553.

[6] N. H. Meyer, K. Zangger, Enhancing the resolution of multi-dimensional heteronuclear NMR spectra of intrinsically disordered proteins by homonuclear broadband decoupling, Chem. Commun. 50, (2014), 1488–1490.

[7] L. Castañar, T. Parella, Broadband 1H homodecoupled NMR experiments: recent developments, methods and applications, Magn. Reson. Chem. 53, (2015), 399-426.

[8] M. Foroozandeh, P. Giraudeau, D. Jeannerat, Broadband 13C-Homodecoupled Heteronuclear Single-Quantum Correlation Nuclear Magnetic Resonance ChemPhysChem 12, (2011), 2409–2411;

[9] L. Castañar, P. Nolis, A. Virgili, T. Parella, Simultaneous Multi-Slice Excitation in Spatially Encoded NMR Experiments, Chem. Eur. J. 19, (2013), 15472–15475.

[10] A. Cotte and D. Jeannerat, 1D NMR Homodecoupled 1H Spectra with Scalar Coupling Constants from 2D NemoZS-DIAG Experiments, Angew.Chem.Int. Ed. 54, (2015), 6016–6018

[11] J. A. Aguilar, S. Faulkner, M. Nilsson, G. A. Morris, Pure Shift 1H NMR: A Resolution of the Resolution Problem? Angew. Chem. Int. Ed. 49, (2010), 3901–3903.

[12] G. A. Morris, J. A. Aguilar, R. Evans, S. Haiber, M. Nilsson, True Chemical Shift Correlation Maps: A TOCSY Experiment with Pure Shifts in Both Dimensions, J. Am. Chem. Soc. 132, (2010), 12770–12772.

[13] Juan A. Aguilar, A. A. Colbourne, Julia Cassani, M. Nilsson, Gareth A. Morris, Decoupling Two-Dimensional NMR Spectroscopy in Both Dimensions: Pure Shift NOESY and COSY, Angew. Chem. 124, (2012), 6566–6569.

[14] N. Giraud, L. Béguin, J. Courtieu and D. Merlet, Nuclear Magnetic Resonance Using a Spatial Frequency Encoding: Application to J-Edited Spectroscopy along the Sample, Angew. Chem., Int. Ed., 49, (2010), 3481-3484; Angew. Chem., 122, (2010), 3559-3562.

[15] N. Giraud, D. Pitoux, J. M. Ouvrard and D. Merlet, Combining J-Edited and Correlation Spectroscopies Within a Multi-dimensional Spatial Frequency Encoding: Toward Fully Resolved 1H NMR Spectra, Chem.–Eur. J., 19, (2013), 12221-12224.

[16] D. Pitoux, B. Plainchont, D. Merlet, Z. Y. Hu, D. Bonnaffé, J. Farjon and N. Giraud, Fully Resolved NMR Correlation Spectroscopy, Chem.–Eur. J., 21, (2015), 9044-9047.

[17] Sandeep Kumar Mishra, N. Lokesh and N. Suryaprakash, Clean G-SERF an NMR experiment for the complete eradication of axial peaks and undesired couplings from the complex spectrum, RSC Adv., 7, (2017), 735-741.

[18] L. Frydman, T. Scherf and A. Lupulescu, The Acquisition of Multidimensional NMR Spectra Within a Single Scan, Proc. Natl. Acad. Sci. U. S. A., 99, (2002), 15858–15862.

[19] A. Tal and L. Frydman, Single-Scan Multidimensional Magnetic Resonance, Prog. Nucl. Magn. Reson. Spectrosc., 57, (2010), 241–292

[20] P. Nolis, M. Pérez-Trujillo and T. Parella, Multiple FID Acquisition of Complementary HMBC Data, Angew. Chem., Int. Ed., 46, (2007), 7495–7497.

[21] David M. Parish and Thomas Szyperski, Simultaneously Cycled NMR Spectroscopy, J. Am. Chem. Soc, 130, (2008), 4925-4933

[22] T. Fäcke and S. Berger, SERF, a New Method for H, H Spin-Coupling Measurement in Organic Chemistry, J. Magn. Reson., Ser. A, 113, (1995), 114-116.

[23] R. Brüschweiler, J. C. Madsen, C. Griesinger, O. W. SØrensen, R. R. Ernst, Two-dimensional NMR spectroscopy with soft pulses, J. Magn. Reson. 73, (1987), 380–385.

[24] L. Emsley, G. Bodenhausen, Self-refocusing effect of 270° Gaussian pulses, Applications to selective two-dimensional exchange spectroscopy, J. Magn. Reson., 82, (1989), 211-221.

[25] J. Cavanagh, J. P. Waltho, J. Keeler, Semiselective two-dimensional NMR experiments, J. Magn. Reson., 74, (1987), 386-393.

[26] H. Oschkinat, A. Pastore, P. Pfändler, G. Bodenhausen, Two-dimensional correlation of directly and remotely connected transitions by z-filtered COSY, J. Magn. Reson., 69, (1986), 559-566

[27] Y. S. Jadeja, K. l. M. Kapadiya, H. J. Jebaliya, A. K. Shah, and R. C. Khunt, Dihedral angle study in Hesperidin using NMR Spectroscopy, Magn. Reson. Chem 55, (2017), 589–594

[28] J. L. Nieto, A. M. Gutierrez, ^{1}H NMR Spectra at 360 MHz of Diosmin and Hesperidin in DMSO solution, Spectroscopy Letters, 19, (1986) 427-434.

[29] D. Sinnaeve, M. Foroozandeh, M. Nilsson, G. A. Morris, A General Method for Extracting Individual Coupling Constants from Crowded ^{1}H NMR Spectra, Angew. Chem. Int. Ed., 54, (2015), 1 – 5.

[30] J. Mauhart, S. Glanzer, P. Sakhaii, W. Bermel, K. Zangger, Faster and cleaner real-time pure shift NMR experiments, J. Magn. Reson., 259, (2015), 207-215.

Chapter 3

DQF *J*-RES NMR: Suppressing the singlet signals for improving the *J*-RES spectra from complex mixtures.

1. Introduction

J-RES spectroscopy finds widespread use in chemical, biological, medical, and environmental studies. This method finds routine use in chemistry for measurement of accurate *J*-couplings and also for resolving peaks in crowded spectral regions. Although one-dimensional (1D) ^{1}H NMR spectroscopy is the most reliable technique in metabolomics due to the rapid acquisition and straightforward quantitative aspects; nevertheless, severe spectral overlap of the peaks is a bottleneck to the identification and quantification of a large number of metabolites. *J*-RES NMR overcomes this spectral congestion greatly by spreading the chemical shift and multiplet information along two orthogonal frequency axes. As a result, *J*-RES NMR spectra have found widespread use in NMR based metabolomics for metabolic profiling studies of biofluids[1–14], like human urine[2] plasma[3,9], cerebrospinal fluids[4]. Diverse studies on plants[10], fish[11], beer [12], tissues [14], kinetic drug metabolism[13], etc. have been carried out by *J*-RES NMR spectroscopy. All such studies have highlighted the advantage of improved peak dispersion in *J*-RES NMR by reducing peak overlaps in highly overlapped regions[1,7]. Improved peak dispersion in projected *J*-RES spectra has been demonstrated to enhance the quality and interpretations of multivariate models such as PCA (Principal Component Analysis) and statistical correla-tion analyses such as STOCSY (statistical total correlation spec-troscopy)[1,7]. *J*-RES spectroscopy has also been used to distinguish enantiomers in chiral aligned media[15–17], to extract one-bond heteronuclear couplings in large macromolecules[18,19], and to improve resolution in DOSY-type experiments[20]. However, the full potential of the *J*-RES technique remains underexploited. The dispersive phase twisted lineshape inherent to the technique renders absorption mode representation of the spectrum ineffective. As a result, only the absolute value of the complex spectra are in routine use. Since the dispersive Lorentzians are broader compared to absorptive Lorentzians, the absolute value mode 2D *J*-RES spectrum in general display broad peaks sacrificing much of the potential high resolution that could be achieved if the dispersive contribution could be suppressed. The problem gets further aggravated in complex mixtures where the concentration of certain metabolites is much higher than the rest. Thus, the broad dispersive tails from the strong peaks obscure the low-intensity peaks in the neighborhood. This often leads to ambiguity in the assignment of metabolite peaks.

Several attempts have been made over the years for producing absorption mode 2D *J*-RES spectra- modified pulse sequences to suppress the dispersive terms[21–23], combined data manipula- tion and pattern recognition[22], purely data post-processing[24], adiabatic z-filtered *J*-RES spectroscopy with multiplet reduction algorithm [25], and recent absorptive *J*-RES spectroscopy by using homo-decoupling schemes such as BASHD (Band Selective Homodecoupling)[26], PSYCHE (Pure Shift Yielded by Chirp Excitation) [27], ZS (Zangger and Sterk)[21] and BIRD (Bilinear Rotation Decoupling)[28]. These approaches have not found routine use as either lower sensitivity (10–20% of regular proton sensitivity at best), complicated optimization and post-processing, and longer experimental time hampers their efficiency. In overall, the regular magnitude mode *J*-resolved technique is extensively used in all applications due to higher sensitivity, better resolution compared to 1D ^{1}H NMR, and simpler execution features.

In the present work, we overcome one key limitation of regular magnitude mode *J*-RES NMR without attempting absorptive mode representation and without losing much of its sensitivity. In metabolomics as well as other applications that involve complex mixtures, quite often some components of the mixture display very high-intensity singlet peaks owing to their high abundance in the mixture and a single spectral line instead of a multiplet. This is particularly observed for complex mixtures such as human urine (where the creatinine singlet peaks generally dominate the whole ^{1}H 1D spectrum) and many other samples. The whole ^{1}H 1D spec- trum is dominated by the intense singlet peaks resulting in the lower intensity of the weaker signals present in the spectrum, and the same is true of 2D *J*-RES spectrum. Any weak intensity multiplet peak close to the intense singlet signals gets obscured in the 2D *J*-RES spectrum. While presaturation at multifrequency or WET technique[29] can be used to suppress such multiple intense singlets, that way one will also eliminate any important multiplet peaks which are eclipsed by the strong and broad singlet resonances. In the present work, we show that a double-quantum filter (DQF)[30–35] when combined with regular magnitude mode *J*-RES technique can greatly reduce the intensity of the singlet resonances leading to better resolution of any weak multiplet peak which is either eclipsed or adjacent to the intense singlet metabolite peaks. High-quality *J*-RES spectrum from three samples- a mixture of eight compounds, human urine, and a plant extract are reported. It is worthwhile to mention that the double- quantum filter has also been used in COSY-DOSY experiment to increase the validity of monoexponential fits for the determination of diffusion coefficients in complex mixtures[36]. Recently, the intermolecular double-quantum coherence based *J*-RES spectroscopy has also been demonstrated for overcoming line broaden- ing induced by variations in susceptibility in intact biological tissue samples. This method utilizes the fact that intermolecular multiple-quantum coherences arising from distant dipolar interactions in such semi-solid samples do not suffer from the field inhomogeneities[37,38].

2. DQF *J*-RES pulse sequence

Fig. 3.1 displays the pulse sequence for acquiring Double Quantum Filtered (DQF) 2D J-RES spectrum. It involves the application of a DQ filter before the evolution period of the J-RES sequence to eliminate the intense singlet signals. The narrow rectangular bars are nonselective 90° pulses while broad rectangular bars are refocusing nonselective 180° pulses. CW represents continuous wave for pre-saturation of water signal during recycle delay. The first gradient G_5 dephases any components of transverse magnetization which may be present due to the previous scans and allows only longitudinal magnetization (along the z-axis) to take part in the pulse sequence. The first 90° pulse between the time points 'a' and 'b' creates the signals denoted by I_{1y}, I_{2y}, I_{3y} at time point 'b.' During time period 2D from 'b' to 'c' these signals evolve into antiphase coherences such as $I_{1x}I_{2z}$, $I_{2x}I_{1z}$, $I_{3x}I_{4z}$ under homonuclear scalar couplings to adjacent signals. The second 90° pulse between time points 'c' and 'd' creates all kinds of multiple-quantum coherences such as $I_{1x}I_{2y}$, $I_{2x}I_{1y}$, $I_{3x}I_{4y}$. The third 90° pulse between time points 'e' and 'f' flanked by gradients in the ratio of G_1:$2G_1$ (coher- ence selection gradients) are used to select only double-quantum coherences present between time points 'd' and 'e' as in DQF-COSY. This suppresses the strong singlet resonances from the final spectrum. Again during the time period 2D from time point 'f' to 'g' (as Δ_1 is a small delay of 1 ms) the signals evolve under scalar coupling and refocus to the initial terms I_{1y}, I_{2y}, I_{3y} etc. at time point 'g.' Due to the difference in the magnitude of scalar couplings between different signals all antiphase coherences does not get refocused to inphase coherences. Hence, the fourth 90° pulse between time points 'g' and 'h' creates the terms $2I_{1x}I_{2y}$ and I_{1z} (and similar terms for other signals) from the antiphase and desired inphase coherences respectively at time point 'h.' The gradient G_0 between time points 'h' and 'i' is used for dephasing all transverse components of magnetization. The final terms left at time point 'i' are Z-spin orders such as I_{1z}, I_{2z}, I_{3z}, etc. A zero-

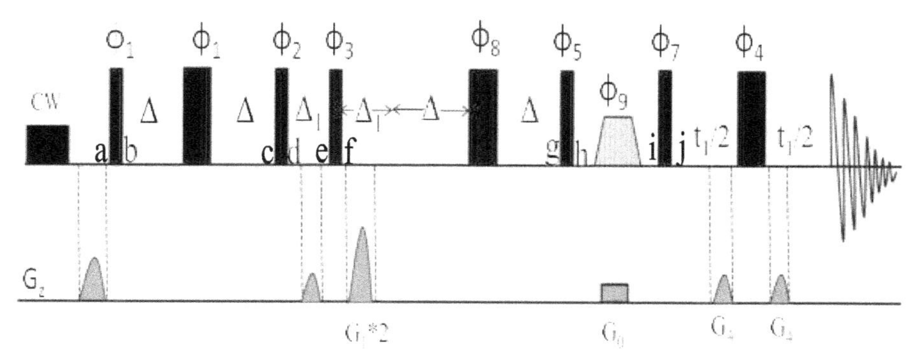

Fig. 3.1. The pulse sequence for DQF *J*-RES spectroscopy. The narrow rectangular bars are nonselective 90° pulses while broad rectangular bars are refocusing nonselective 180° pulses. CW represents continuous wave for pre-saturation of water signal during recycle delay. The delay 2D is tuned to 40 ms. G_5 (=30.5G/cm) and G_0 (=25.1G/cm or 0.8G/cm for ZQ filter) are dephasing gradient. G_1 (22.7G/cm) and G_1*2 are used to select the DQ coherences. G_4 (=8.3G/cm) flanked by the refocusing pulses are coherence selection gradient for the signals that go through an echo. A zero-quantum filter can also be applied between time points 'h' and 'i'. However, this is optional for magnitude mode *J*-RES spectroscopy as any residual signal via ZQ pathway does not interfere with the multiplets generated by t_1 modulation of the scalar coupling interaction.

quantum filter can also be applied between time points 'h' and 'i' shown with trapezoid shape to improve the spectrum. However, this is optional as any residual signal via ZQ pathway does not interfere with the multiplets in a magnitude mode *J*-RES spectroscopy. The last 90° pulse is applied between time points 'i' and 'j' which starts the same time events as in regular *J*-resolved sequence. The nonselective refocusing pulse flanked by the G_4 gradients at the centre of the t_1 dimension ensures the evolution of only homonuclear scalar couplings leading to the final terms as I_{1y} and $-2I_{1x}I_{2z}$ and similarly for other signals such as I_2, I_3, etc. Low power CW presaturation was also applied during the D delays to suppress the water signals.

3. Experimental

3.1. Sample preparation and data acquisition

All the one and two-dimensional experiments reported in this paper were carried out on an 800 MHz NMR spectrometer equipped with a CPTCI cryoprobe with z-axis gradient only at 300 K temperature. All *J*-resolved spectra were acquired and pro- cessed in magnitude mode. Three samples were targeted for evaluating the performance of our new technique- a mixture of eight compounds as detailed below, one human urine sample, and an extract of G. Pedunculata dried fruit which represents complex metabolomics mixture.

3.2. A complex mixture of eight metabolites

A complex mixture designated as 1 was prepared by mixing the following compounds- valine, tyrosine, phenylalanine, tryptophan, betaine, creatinine, cysteine, and acetylacetone in 500 mL of D_2O and 500 mL of phosphate buffer (pH = 7.0). The final concentration of the. Finally solution was: 8.5 mM of valine, 4.9 mM of tyrosine, 7.8 mM of phenylalanine, 24 mM tryptophan, 83 mM for betaine, 62 mM for creatine, 9 mM for cysteine and 49 mM for acetylacetone, the 500 mL solution was transferred to a 5 mm NMR tube for NMR experiment. Regular 2D *J*-RES spectrum displayed in Fig. 3.2a) was recorded on this mixture with the following acquisition and processing parameters: t_1 and t_2 acquisition times of 392 ms and 369 ms respectively, spectral width 8333 Hz in F_2, and 48 Hz in F_1 dimension, number of transients was 16. Zero-filled to 16,384 and 128 data points in F2 and F_1 dimension respectively. Processed with unshifted sine bell window in both F_2 and F_1 dimensions in magnitude mode. DQF-*J*RES spectrum shown in Fig functions. 3.2b was acquired with the same acquisition, and pro- cessing parameters as above except the number of transients were 32 per increment. Total experimental time was 29 min 48 s for magnitude mode 2D *J*-RES experiment and 1 hr and 1 min for DQF-*J*-RES experiment. The 2D delay was tuned to 40 ms. Gradients used in regular 2D *J*-RES experiment were 15% (=8.3G/cm) before and after the refocusing pulse with shape SINE.100. Gradients used in DQF *J*-RES NMR are G_1 = 42.5% (=22.7G/cm), G_0 = 47% (=25.1G/cm) or 1.5% (=0.8G/cm for ZQ filter), G_4 = 15% (=8.3G/cm) and G_5 = 57% (=30.5G/cm).

3.3. Dried fruit extract of G. Pedunculata

The dried, ground fruit of Garcinia Pedunculata weighing 5 g was subjected to extraction in a Soxhlet apparatus using 50 mL of non- polar solvent n-hexane for 12 h. The solvent was evaporated by rotary evaporator. 500 mL of $CDCl_3$ was added to the dried sample, and this solution was taken into a 5 mm NMR tube for NMR experiment. Regular 2D *J*-RES spectrum of dried fruit extract displayed in Fig. 3.4(a), (c), (e), and (g) was acquired with t_1 and t_2 acquisition times of 392 ms and 428 ms respectively, with a spectral width of 8333 Hz in F_2-dimension and 70 Hz in F_1-dimension, number of transients was 32. The spectrum was zero-filled by 16,384 and 128 data points in F_2 and F_1 dimensions respectively. The spectrum was processed with unshifted sine bell window functions in both F_2 and F_1 dimensions in magnitude mode. DQF *J*-RES spectrum dis- played in Fig. 3.4b, d, f, and h was acquired with the same parameters as mentioned above except the number of transients were 64 per increment and 2Δ delay of 40 ms. Total experimental time was 1 h and 31 min for magnitude mode 2D *J*-RES experiment and 3 h and 11 min for Double Quantum Filtered 2D *J*-RES spectrum. Same gradients were used as in the DQF *J*-RES sequence mentioned above.

3.4. Lyophilized human urine sample

Human urine from a volunteer was collected and kept at 80 °C deep freezer. Approval was taken for NMR method development on bio-fluids as per Institute Ethics Committee (Letter No. B17/CBMR/ IEC/EMP/5/2017). The sample was thawed, and 1 mL of the urine sample was taken for further processing. The sample was centrifuged at 12,000g for 5 min, and the supernatant was collected. The supernatant was frozen at 80 °C again and put it into Lyophi-lizer for getting the powder form. The powder was mixed with 260 mL D_2O and 260 mL Buffer of pH = 7.4. The solution was centrifuged at 12,000g for 5 min. 500 Ml Solution was taken into 5 mm NMR tube for NMR experiments. Regular 2D *J*-RES spectrum of Lyophilized human urine shown in Fig. 3.6(a), (c), (e), (g), (i) and (k) was acquired with t_1 and t_2 acquisition times of 392 ms and 369 ms respectively, with spectral width of 8333 Hz in

F_2, and 65 Hz in the F_1 dimension, the number of transients was 64. Zero-filled to 16,384 and 128 data points in F_2 and F_1 dimension respectively. Regular 2D *J*-RES spectrum was processed with unshifted sine bell window functions in both F_2 and F_1 dimensions in magnitude mode. DQF 2D *J*-RES spectral portions displayed in Fig. 3.6(b), (d), (f), (h), (j) and (l) was acquired with the same acquisiton and processing parameters as mentioned above. Total experimental time was 1 h and 57 min for regular *J*-RES experiment and 2 h and 4 min for DQF *J*-RES. Same gradients were used as in the DQF *J*-RES sequence mentioned above, and 2Δ delay of 40 ms was used.

4. Results and discussion

4.1. Application to a complex mixture of eight metabolites

The 45° tilted magnitude mode *J*-RES spectra of the mixture 1 are displayed in Fig. 3.2(a) and (b) from regular *J*-RES sequence and DQF *J*-RES sequence respectively. Small portions of the spectra marked with P, Q, R, and V from (3.2a) (from regular *J*-RES) are shown expanded in Fig. 3.2(c) and (e) respectively with dotted boxes. The same portions in (3.2b) (from DQF *J*-RES) are marked with S, T, U, and W and shown expanded in Fig. 3.2(d) and (f) respectively. Comparisons of the spectral portion R vs. U (in 3.2c vs. 3.2d)), and also V vs W (in 3.2e vs 3.2f) reveals the superior quality spectrum from DQF *J*-RES sequence- the tyrosine and valine multiplets along F_1 are obscured in the regular *J*-RES spectrum due to overlap with the creatine and acetylacetone strong singlet peaks respectively, but get well resolved in DQF *J*-RES spectrum due to the efficient suppression of the strong singlet peaks. Assignment of the peaks are shown in (3.2d) and (3.2f). Inspection of the spectral region inside box U in 3.2d reveals the significantly enhanced intensity of the tyrosine multiplet relative to the creatine singlet. However, the singlet peak from creatine methyl peak in the box R in (3.2c) is very high in intensity compared to the overlapped multiplet from tyrosine germinal proton as the concentration of creatine in the mixture were almost 12 times higher than tyrosine. Similarly the acetylacetone singlet peak in the box W (in 3.2f)) becomes very weak relative to the valine multiplet peak after the application of DQF. In contrast, the same acetylacetone singlet peak in box V in regular *J*-RES spectral portion is quite intense relative to the valine multiplet as the concentration of the former was 6 times higher than the later. Comparison of the spectral region inside dotted box portions P vs. S shows suppression of strong singlet peak in DQF *J*-RES spectrum and hence more clarity of the two doublet of doublets (from Tryptophan and phenylalanine showed in dotted box S) in the neighborhood. A similar comparison of the peaks inside dotted boxes Q and T shows that the two doublet of doublets from tyrosine and phenylalanine are better read out from DQF *J*-RES spectrum. In regular *J*-RES spectrum, the strong tail from the singlet at zero frequency along F_1 can lead to ambiguity in assignment when a multiplet component is buried inside it.

Suppression of singlet intensity to almost null clears such ambiguity. Comparison of the 1D projections from 3.2c and 3.2d are shown in 3.2c' and 3.2d' respectively for the comparable intensity of the larger signals. Similarly, 3.2e and 3.2f are shown in 3.2e' and 3.2f' respectively. Comparison of these projections reveals improved visibility of the multiplet peaks marked with a star in the DQF projections relative to the singlet peaks (without star) in the regular *J*-RES projections. The receiver gain value got optimized to 20.3 in (a, c, e) in regular *J*-RES, a small value for the observation of the smaller signals. In contrast, the DQF *J*-RES optimized this value to 203 (in b, d, and f) which improved the visibility of the smaller signals. Thus, the singlet peaks combined with the higher concentration of that metabolite leads to severe spectral readout problem for the weaker peaks due to their order of magnitude higher intensity than the weaker multiplets from the less abundant metabolites. The DQF filter scales down all singlet peaks (more than the multiplet peaks) revealing the weaker multiplet structures clearly. Although on the absolute S:N scale (as is shown in next section Fig. 3.3) the DQF *J*-RES is a poor spectrum, however, suppressing the intense signals relative to the weaker signal has improved the assignment of the weaker multiplet peaks in DQF *J*-RES spectrum. Regular ^{1}H NMR with presaturation, regular *J*-RES projection, and DQF *J*-RES projection are compared for the same intensity of the most intense singlet peak in Fig. 3.3a, b, and c respectively from the AA's mixture. This comparison shows a few intense singlet signals marked with a star (at 3.24, 3.0, 2.26, ppm) dominate the regular ^{1}H NMR (a) and regular *J*-RES projection (b). In contrast, DQF *J*-RES projection in c shows great improvement in the intensity of the smaller signals relative to the curtailed singlet signals marked with a star. Further, we made a comparison of the regular *J*-RES projection and DQF *J*-RES projection on the same noise level in (d) and (e) respectively which shows much higher signal to noise ratio (S:N) of the regular *J*-RES projection. Measurement of S:N for four peaks reported in Table 3.1 revealed regular *J*-RES has on average much higher S:N relative to the DQF *J*-RES projection. However, the suppression of the intense singlet peaks improves the visibility of the weak intensity metabolite peaks as the receiver gain gets optimized for the smaller signals in DQF *J*-RES spectrum. This is clearer from the analysis of the 2D spectral portions in Fig. 3.2c-f. Identification of metabolites from

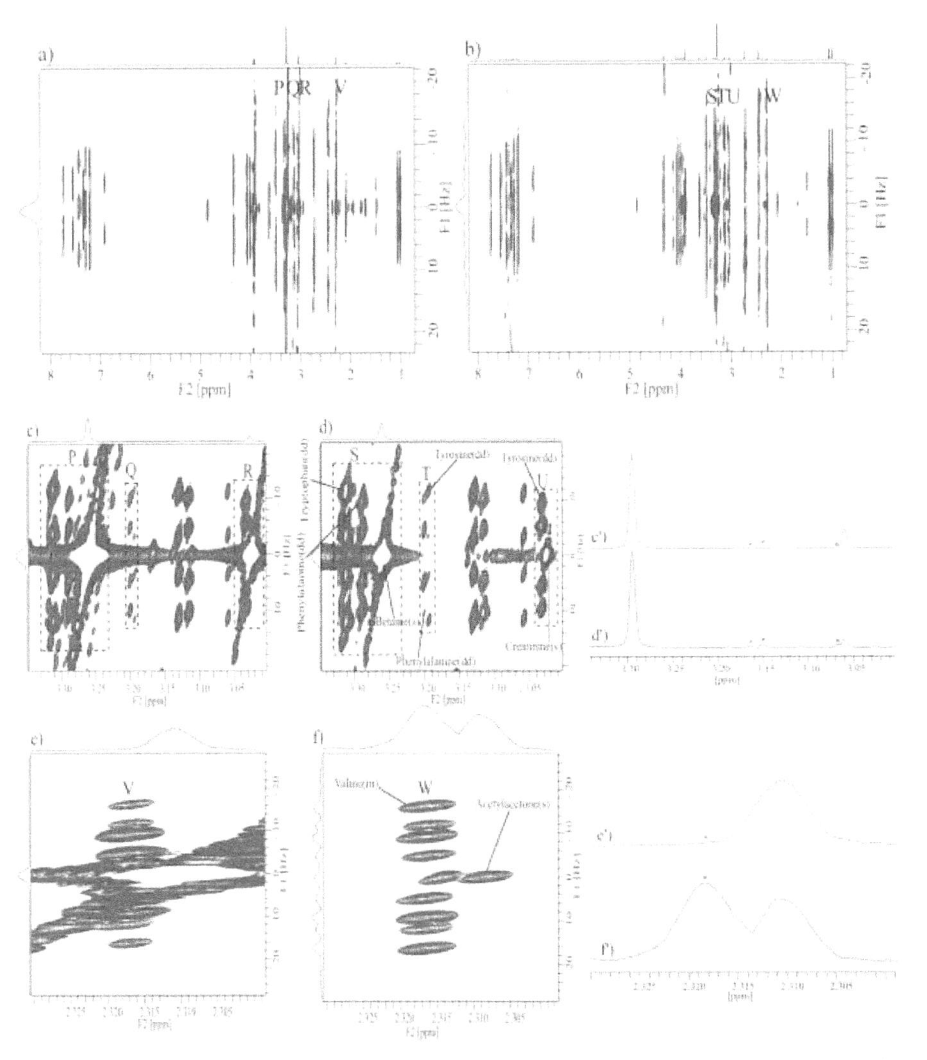

Fig. 3.2. (a) and (b) are the 45° tilted magnitude mode *J*-RES spectrum and DQF *J*-RES spectrum of the mixture 1 respectively. Small portions of the spectra marked with P, Q, R, and V in (a) are shown expanded in

Fig. 3.2(**c**) and (e) respectively with dotted boxes. The same portions in (b) are marked with S, T, U, and W and shown expanded in Fig. 3.2(d) and (f) respectively. Panel 3.2c shows intense singlet peak and its tail along $F_1 = 0$ obscuring the multiplet information inside the dotted boxes P, Q, and R. Panel 3.2d shows that the intensity of the singlet peaks along $F_1 = 0$ is reduced and the multiplet information such as dd of tyrosine inside dotted box U is revealed clearly. The clarity of the multiplets inside the dotted boxes S and T also improves. Comparison of the 1D projections from the 2D portions (c) and (d) are shown in (c') and (d') respectively which reveals improved visibility of the multiplets in DQF projection (shown with a star mark) in d') relative to the singlets (without star mark). Panel (3.2e) from regular *J*-RES 2D portion V shows a multiplet overlapped with an intense singlet which does not allow extraction of the *J*-coupling information. Panel (3.2f) from DQF *J*-RES improves the intensity of the multiplet relative to the singlet and the ddd pattern from valine can be clearly assigned. Comparison of the projections (e') and (f') from (e) and (f) also reveals better visibility of the ddd peak (star mark) of valine relative to the singlets. Thus, reduced dynamic range issues in DQF *J*-RES improves assignment.

bio-fluids is routinely per- formed in metabolomics which often aims for identifying potential biomarkers for diseases. However, severe spectral overlap in the ^{1}H NMR

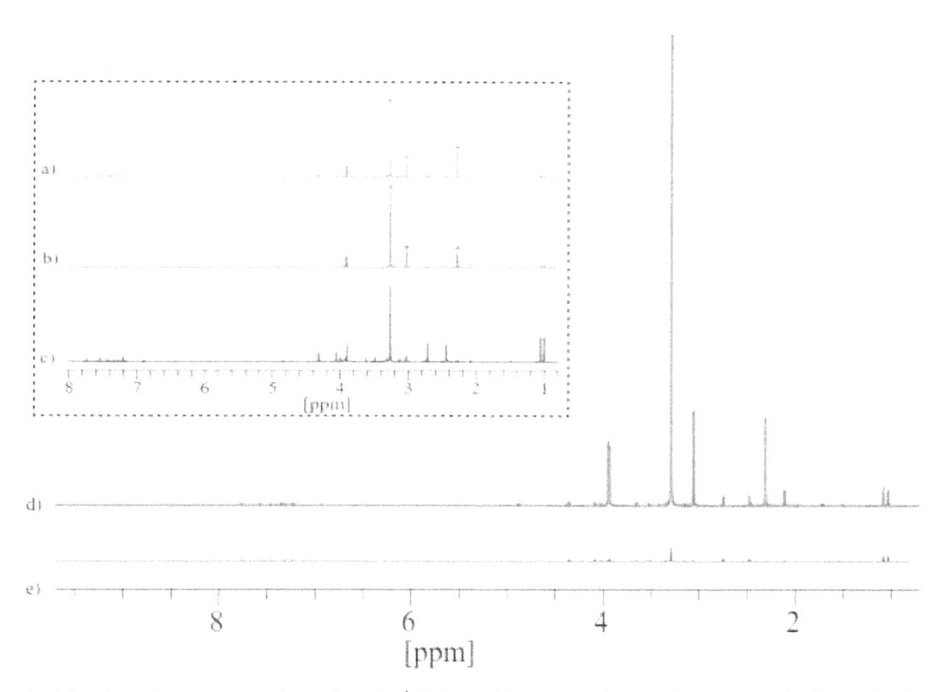

Fig. 3.3. a, b, and c are a comparison of regular ^{1}H NMR with presaturation, regular *J*-RES projection, and DQF *J*-RES projection respectively from the amino acid mixture compared for the same intensity of the most intense

singlet peak. The comparison reveals few intense singlet signals marked with a star (at 3.24, 3.0, 2.26, ppm) dominate the regular ^{1}H NMR (a) and regular *J*-RES projection (b). In contrast, DQF *J*-RES projection in c shows great improvement in the intensity of the smaller signals relative to the curtailed singlet signals marked with a star. (d) and (e) is a comparison on the same noise level of the regular *J*-RES projection and DQF *J*-RES projection, which shows much higher S:N of the regular *J*-RES projection. Measurement of S:N for four peaks reported in Table 3.1 revealed regular *J*-RES has on average much higher S:N relative to the DQF *J*-RES projection despite the later had two times higher transients. The improved receiver gain 203 in (3.3e) relative to 20.3 in regular *J*-RES (3.3d) improves the digitization of the smaller signals compared with the noise. This gain combined with suppression of the intense singlet peaks improves the visibility of the weak intensity metabolite peaks in DQF *J*-RES spectrum. This is also clearer from the analysis of the 2D spectral portions in Fig. 3.2c–f.

spectra from bio-fluid samples hampers the identification of many metabolite signals. 2D NMR such as *J*-RES, TOCSY, HSQC are suitable for reducing spectral overlap. Despite the higher resolution of 2D NMR, it does not get rid of the overlap issue completely in particular for complex mixtures where hundreds of metabolites are present in widely varying concentrations. Thus our improved DQF *J*-RES technique was further evaluated on human urine and plant extract samples.

Table. 3.1. Amino Acids

Note: DQF *J*-RES had two times more number of transients in this case than regular *J*-RES. Therefore, the values inside brackette in the column (A/B) are obtained by multiplying by square root of 2.

Peaks δ(ppm)	1D proj Reg. JRES (S/N) (A)	1D proj DQF JRES (S/N) (B)	(A)/(B)
1.01	72261	27155	2.6(3.7)
7.30	3706	477	7.7(10.8)
7.45	4011	1885	2.1(2.9)
7.55	4217	4559	0.9(1.3)

4.2 Application to a dried fruit extract sample

A methanol extract of G. Pedunculata dried fruit was targeted with DQF *J*-RES sequence, and the resulting spectrum was com- pared with the regular *J*-RES spectrum in Fig. 3.4. Fig. 3.4(a), (c), (e), and (g) are small portions expanded from regular *J*-RES spectrum.

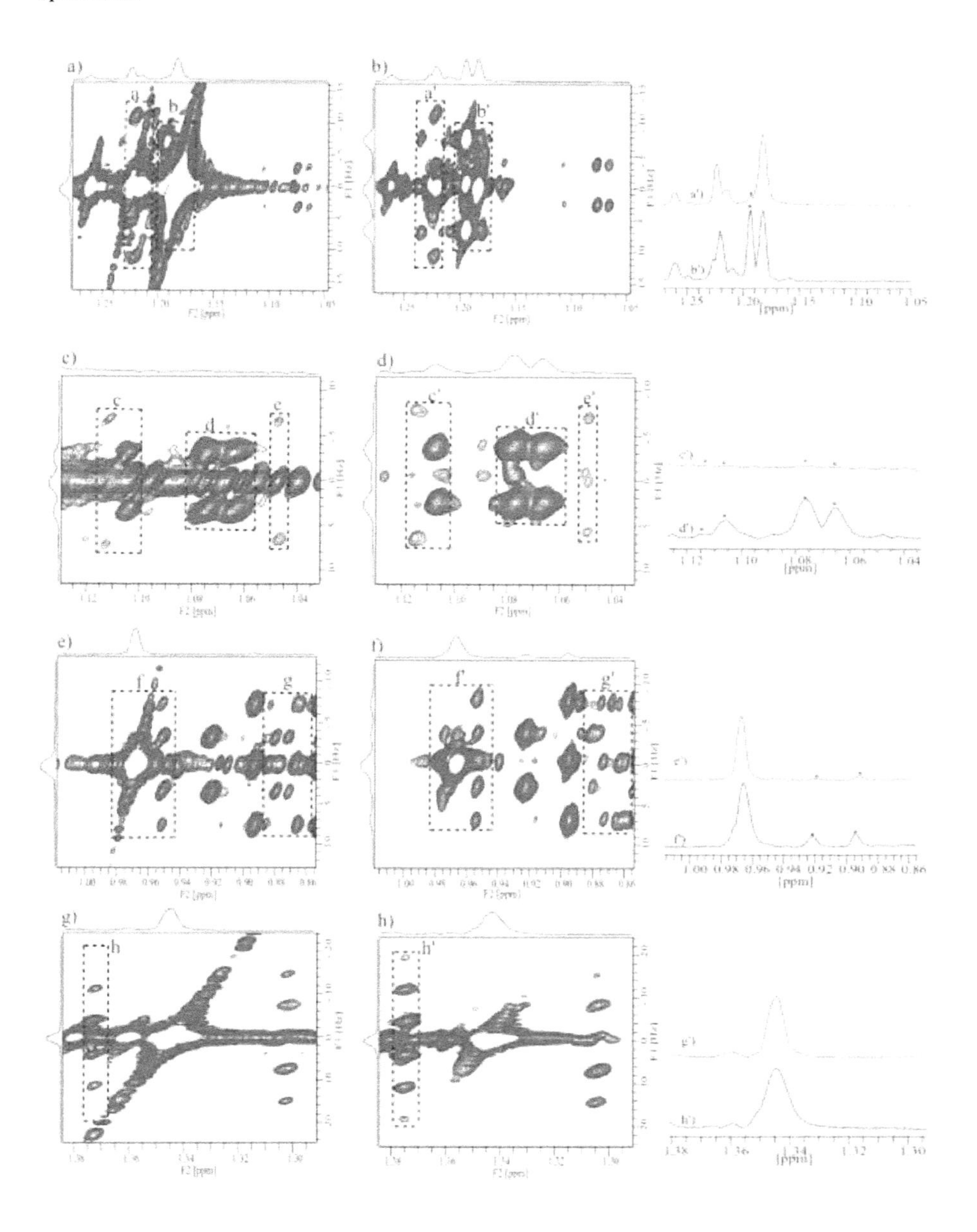

Fig. 3.4. (a), (c), (e), and (g) are small portions expanded from regular *J*-RES spectrum of G. Pedunculata dried fruit extract in methanol. The corresponding regions from the DQF *J*-RES spectrum are shown in Fig. 3.4(b), (d), (f), and (h) respectively. The intense singlet peaks and its tails along $F_1 = 0$ in regular *J*-RES spectrum are greatly reduced in DQF *J*-RES spectral portions in all panels. Comparison of dotted boxes inside the 2D spectral portions (a) and (b), and their projections on right panels (a') vs. (b') shows improved intensity of the multiplets marked with a star in the DQF projection relative to the singlet (shown without star). In (b') the small hump at the left side of the left star marked peak is actually a reduced singlet peak. The strong singlets in 2D portion (a) have much lower intensity in (b). Similarly comparison of the 2D portions (c) and (d) and their projections (c') vs. (d') reveals improved intensity of the multiplets marked with a star in the DQF projection (d'). The absence of the singlets along $F_1 = 0$ is noteworthy in 2D portion (d) relative to (c). Similarly, comparison of (e) vs. (f) and also their projections (e') vs. (f') on right panel confirms improved intensity of the multiplets relative to the singlets (marked with a star) and reduced intensity of singlets along $F_1 = 0$ in 2D portion (f). Similarly, the singlets in (g) are also reduced in (h).

The corresponding regions from the DQF *J*-RES spectrum are shown in Fig. 3.4(b), (d), (f), and (h) respectively. Full spectra are displayed in Fig. 3.11A and 3.11B. Comparison of dotted boxes a vs. a' and b vs. b' from (3.4a) and (3.4b) display better visibility of the multiplets in DQF *J*-RES spectrum in (3.4b) which are otherwise obscured due to the presence of strong singlet peaks and their intense tails along $F_1 = 0$ in (3.4a). Two triplets and one doublet are clearly observed inside dotted box a' and b' in (3.4b).

Comparison of Fig. 3.4(c) and (d) (dotted boxes c vs. c', d vs. d', e vs. e') reveals that regular *J*-RES spectrum is dominated by the intense singlet peaks along $F_1 = 0$ leading to concealing of the multiplet structures from the low abundant metabolites. However, these intense singlet peaks along $F_1 = 0$ get completely suppressed in DQF *J*-RES spectrum in 3.4d) improving the visibility of the multiplets. Similarly, comparison of the Fig. 3.4(e) vs. (f), and (g) vs. (h) reveals efficient suppression of the intense singlet peaks along $F_1 = 0$ in the DQF *J*-RES spectrum improving the clarity of the neighboring multiplet peaks. Comparison of the 1D projections from the compared 2D portions (In left and middle panels) are shown on right panels as (a') vs. (b'), (c') vs. (d'), (e') vs. (f'), g') vs. (h') respectively (plotted for the comparable intensity of the larger signals)- which further confirms improved visibility of the multiplet peaks marked with star in the DQF projections relative to the singlet peaks (without star) in the regular *J*-RES projections. For instance, the star marked triplets in the projection b') have poor visibility in the projection a'). When compared for the same intensity of the most intense singlet peak Fig. 3.5(a) (regular ^{1}H NMR with presaturation), (b) (regular *J*-RES projection), and (c) (DQF *J*-RES projection) reveals that DQF *J*-RES projection in (c) shows great improvement in the intensity of the smaller signals present in the region from 2 to 8 ppm relative to the intense singlet peaks in the region from 0 to 2 ppm. In fact the ^{1}H 1D spectrum in Fig. 3.5(a) is dominated by the singlet marked with a star. Most of these signals (2 to 8 ppm) appear with very poor intensity in regular *J*-RES projection in (b) and in ^{1}H 1D in (a) when compared to the intense singlets in the region from 0 to 2 ppm. Due to the suppression of the intense singlet signals, the receiver gain improved in the DQF *J*-RES spectrum which improved the visibility of the low abundant metabolite peaks which were on the noise level in the regular *J*-RES spectrum. Further, a comparison on the same noise level of the regular *J*-RES projection and DQF *J*-RES projection in (d) and (e) shows much higher S:N of the regular *J*-RES projection. The S:N values for four peaks are reported in Table 3.2 which implies regular *J*-RES has on average much higher S:N relative

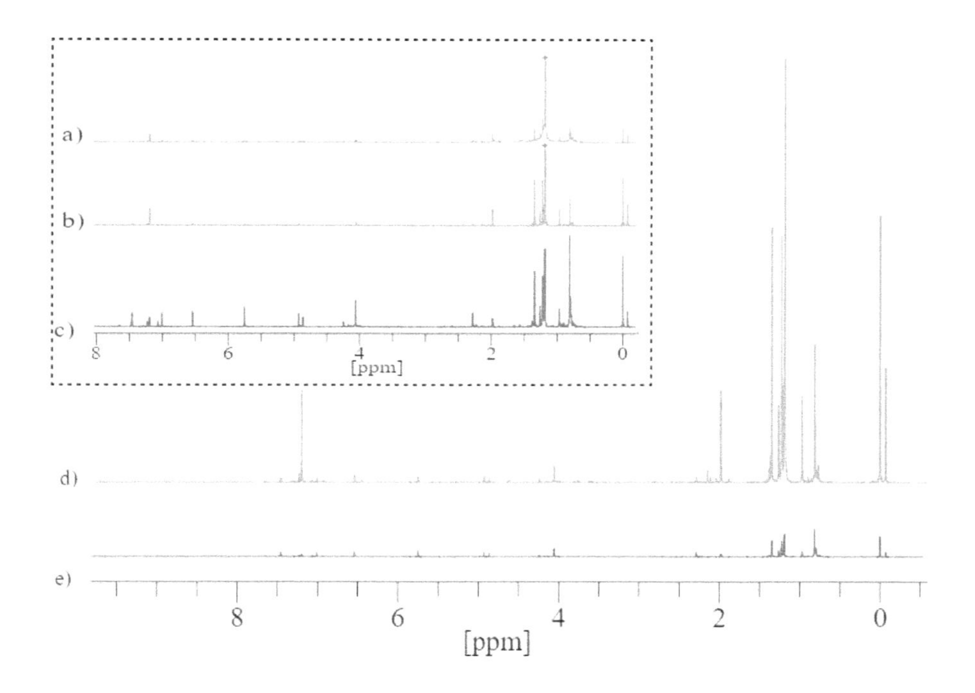

Fig. 3.5. (a)–(c), are the comparison of regular ^{1}H NMR with presaturation, regular *J*-RES projection, and DQF *J*-RES projection respectively from the G. Pedunculata dried fruit extract plotted for the same intensity of the most intense singlet peak marked with a star. The intense

singlet signals marked with star dominate the spectra in (a) and (b) rendering the smaller signals almost undetectable. In contrast, DQF *J*-RES projection in 3.5c shows great improvement in the intensity of the smaller signals relative to the intense singlet signals whose intensity is now substantially reduced relative to the other peaks. Comparison on the same noise level of the regular *J*-RES projection and DQF *J*-RES projection is also shown in (3.5d) and (3.5e) which indicates much higher S:N of the regular *J*-RES projection. table 3.2 with the S:N values for four peaks indicate regular *J*-RES has much higher S:N relative to the DQF *J*-RES projection although DQF *J*-RES, in this case, was recorded with two times more number of transients. However, the suppression of the intense singlet peaks improves the visibility of the weak intensity metabolite peaks as the maximum receiver gain gets optimised for these smaller signals in DQF *J*-RES spectrum (203 relative to 10 in regular *J*-RES spectrum).

to the DQF *J*-RES projection. However, the suppression of the intense singlet peaks improves the visibility of the weak intensity metabolite peaks as the receiver gain gets optimised for the smaller signals in DQF *J*-RES spectrum to 203 (relative to 10 in regular *J*-RES spectrum). This is clearer from the analysis of the 2D spectral portions in Fig. 3.4a-f.

Table. 3.2. Plant Extract

Note: DQF *J*-RES had two times more number of transients in this case than regular *J*-RES. Therefore, the values inside brackette in the column (A/B) are obtained by multiplying by square root of 2.

Peaks δ(ppm)	1D proj Reg. JRES (S/N) (A)	1D proj DQF JRES (S/N) (B)	(A)/(B)
1.55	809	585	1.4(1.9)
1.64	301	568	0.5(0.7)
5.46	534	94	5.6(7.9)
6.99	2313	1508	1.5(2.1)

4.3 Application to a lyophilized human urine sample

The performance of the DQF *J*-RES experiment was further evaluated on a lyophilized human urine sample in D_2O. While full spec- tra are reported in the Fig. S3 3.10A and S3 3.10B, representative 2D *J*-RES spectral sections are shown expanded in Fig. 3.6(a), (c), (e), (g), (i), and (k) from the regular *J*-RES sequence and in (3.6b), (3.6d), (3.6f), (3.6h), (3.6j), and (3.6l) from the DQF *J*-RES sequence. In Fig. 3.6a), close to the right side of the dotted line b an intense singlet is present which does not allow any weak intensity multiplets in its vicinity to be noticeable. For example, there are multiplets along the dotted line b and c but are not detectable due to the intense singlet in the middle of these two lines. Suppression of this intense singlet in (3.6b) allows the multiplets b', c', and d' to be better resolved and their multiplet pattern gets established. While the doublet f is very weak in (3.6a), its intensity improves in (3.6b). Similarly, the doublet of doublet e' displays more intensity in 3.6b) relative to e in (3.6a). However, doublet a in (3.6a) is missing in (3.6b) indicated by the empty dotted box a'. Similarly, the intense singlet from creatinine overlapping the line a in (3.6c) is almost completely suppressed in (3.6d). As a result, the two doublet of doublets marked with a' and b' are clearly observed in (3.6d) whereas in (3.6c) these peaks get completely obscured and lost inside the intense peak.

The appearance of the multiplet c' is also improved in (3.6d). Similarly, the multiplets a' to f' are more clearly detected and established in (3.6f) relative to the peaks a to f in (3.6e) due to the suppression of the intense singlets along $F_1 = 0$ by the DQ filter in (3.6f). Comparison of (3.6g) and (3.6h) reveals that the multiplets within the dotted box are more clearly established in (3.6h) due to the absence of the singlet peaks along $F_1 = 0$. Besides, the multiplet peak b' is much better visible in (3.6h) whereas it is almost buried inside the tail of the singlet from creatinine in (3.6g).

Comparison of the multiplets marked with letters a-o in Fig. 3.6i) with the corresponding peaks in (3.6j) marked with letters a'-o' reveal that the DQF filter effectively suppresses most of the high- intensity singlet signals along $F_1 = 0$ in (3.6j) and improves the clarity of the multiplets compared to (3.6i). Multiplet patterns of a', b', i', g', h', j', n', o' in (3.6j) becomes more clear compared to a, b, i, g, h, j, n, o in 3.6I due to the suppression of the strong singlet signals along $F_1 = 0$. For instance, that c' is a doublet of doublet and h' is a doublet becomes unambiguous in (3.6j) but singlet intensity along $F_1 = 0$ does not allow this to be confirmed in (3.6i). While doublet d and f are present in (3.6i), corresponding peaks d' and f' are missing in (3.6j). Similarly, while the peaks k' and o' are present in (3.6j), corresponding peaks k and o are missing in (3.6i). On fewer occasions a few peaks were found missing in the DQF *J*-RES spectrum, for example, the doublet of doublet b in 3.6 k) is

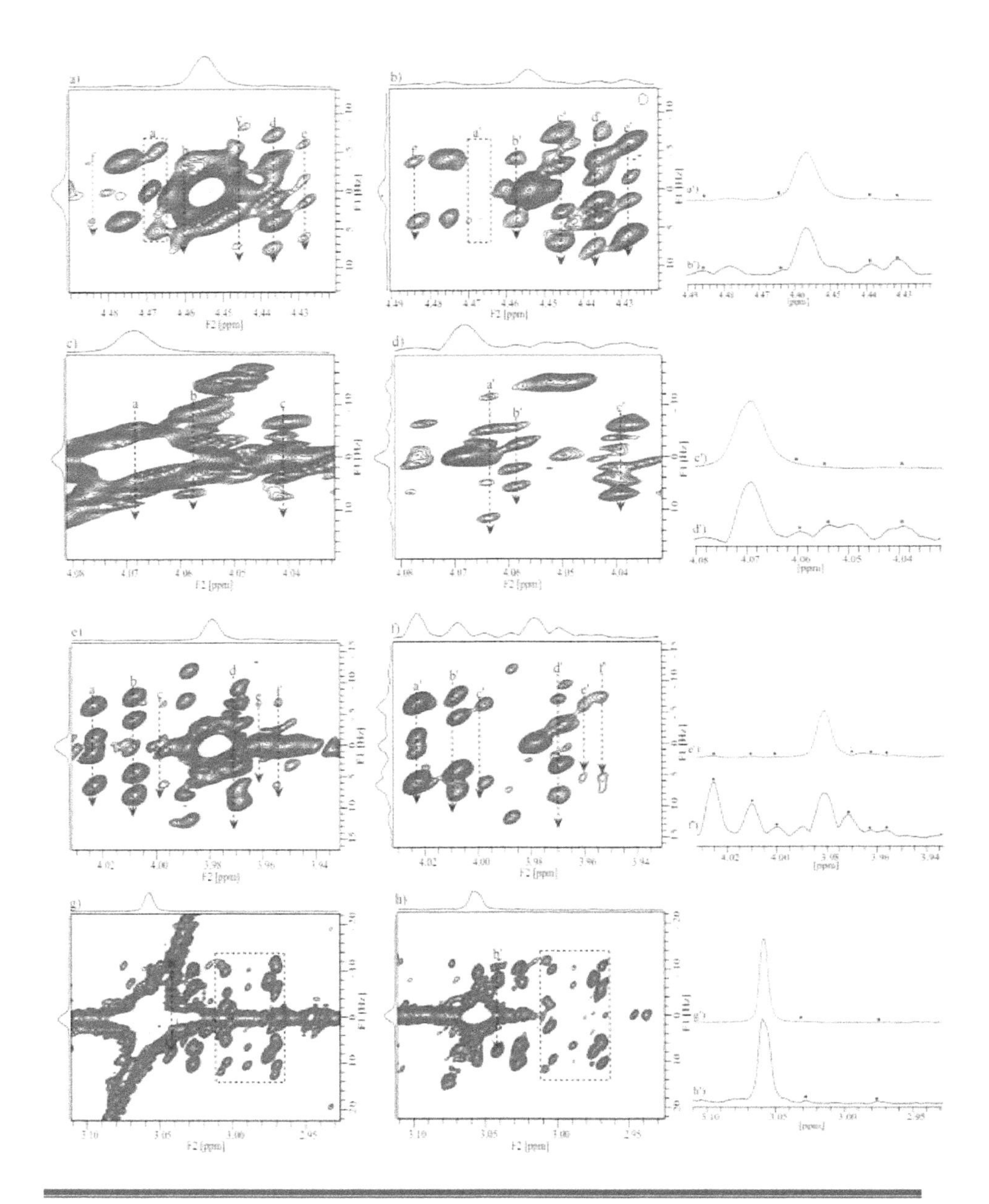

Fig. 3.6. (a), (c), (e), (g), (i), and (k) are small portions expanded from regular *J*-RES spectrum of a lyophilized human urine sample in D_2O. The corresponding regions from the DQF *J*-RES spectrum are shown in (b), (d), (f), (h), (j), and (l) respectively. In all spectral portions from regular *J*-RES spectrum (all left side panels), the intense singlets dominate along $F_1 = 0$ which prevents unambiguous determination of a multiplicity of the signals which are overlapped with the singlets. In contrast the corresponding spectral portions from DQF *J*-RES spectrum (all middle panels) uncovers many hidden multiplets such as b' , c' in (3.6b) vs. b, c in (3.6a); a' , b' in (3.6d) vs. a, b in (3.6c); d' in (3.6f) vs. d in (3.6e); b' in (3.6h) vs. b in (3.6g). In addition, many other peaks have greater clarity in the DQF *J*-RES spectrum as detailed in results and discussions. The right panel (topmost) show the comparison between the projections (a') vs. (b') extracted from the 2D spectral portions on left (a) and (b) respectively. The enhanced intensity of the multiplets in the DQF projection (b') are marked with star relative to the singlet shown without a star. These comparisons are for the same intensity of the singlets between (a') and (b'). Similar results are obtained from the comparison of the other *J*-RES projection (top in each panel) vs. DQF *J*-RES projection (bottom in each panel) viz. (c') vs. (d'); (e') vs. (f'); (g') vs. (h'); (i') vs. (j'); (k') vs. (l').

missing in 3.6 l). This is further displayed for many other peaks in Figures 3.9 and 3.10 for plant extract and lyophilized urine samples respectively. Comparison of the 1D projections from the compared 2D portions on the left and middle panels are shown on right panels a') vs. b'), c') vs. d'), e' vs. f'), g') vs. h'), i') vs. j'), k') vs. l') respectively which further confirms improved visibility of the multiplet peaks relative to the intense singlet peaks. In order to address the disappearance of certain multiplet peaks in DQF *J*-RES spectrum, we investigated the delay-dependent modulation of the signals by the DQF. A series of DQF spectra were recorded for different values of delay Δ (20, 30, 40, and 50 ms) on the same lyophilized human urine sample and reported in Figure 3.8. Since, in a metabolomics complex mixture, large variations exist in *J*-values due to the presence of a large number of metabolites; thus some signals can disappear in the DQF *J*-RES spectrum when their sine modulated antiphase precursor terms after a 2Δ period approach zero values. Therefore, for certain *J*-values the creation of double quantum coherence is maximum; however, for other *J*-values, this could be minimum. This situation changes as a function of the delay Δ. As we show in Figure 3.8, more multiplets can be recovered by a combined analysis of these DQF *J*-RES spectra recorded with different values of Δ and comparing with each other as well as to the regular *J*-RES. This process is although time-consuming can be more fruitful. Comparison of the Fig. 3.7a) (regular ^1H NMR with presaturation), (3.7b) (regular *J*-RES projection), and (3.7c) (DQF *J*-RES projection) reveals that even for the same receiver gain and same acquisition parameters (recycle delay, number of transients were also same in this case, and acquisition time) the DQF *J*-RES projection shows much better visibility of the smaller signals. The intense singlet signals marked with a star at 3.05 ppm and 4.07 ppm (from creatinine) in (3.7a) and (3.7b) dominate the ^{1}H 1D and regular *J*-RES projection respectively. As a result, very few signals in ^{1}H 1D in (3.7a) and regular *J*-RES projection in (3.7b) display intensity comparable to the singlet from creatinine. With the suppression of the singlets in (3.7c), visibility of the weak intensity metabolite peaks are improved. Further, a comparison on the same noise level of the regular *J*-RES projection and DQF *J*-RES projection in (3.7d) and (3.7e) shows much higher S:N of the regular *J*-RES projection. The S:N values for five peaks are reported in Table 3.3 which implies regular *J*-RES has on

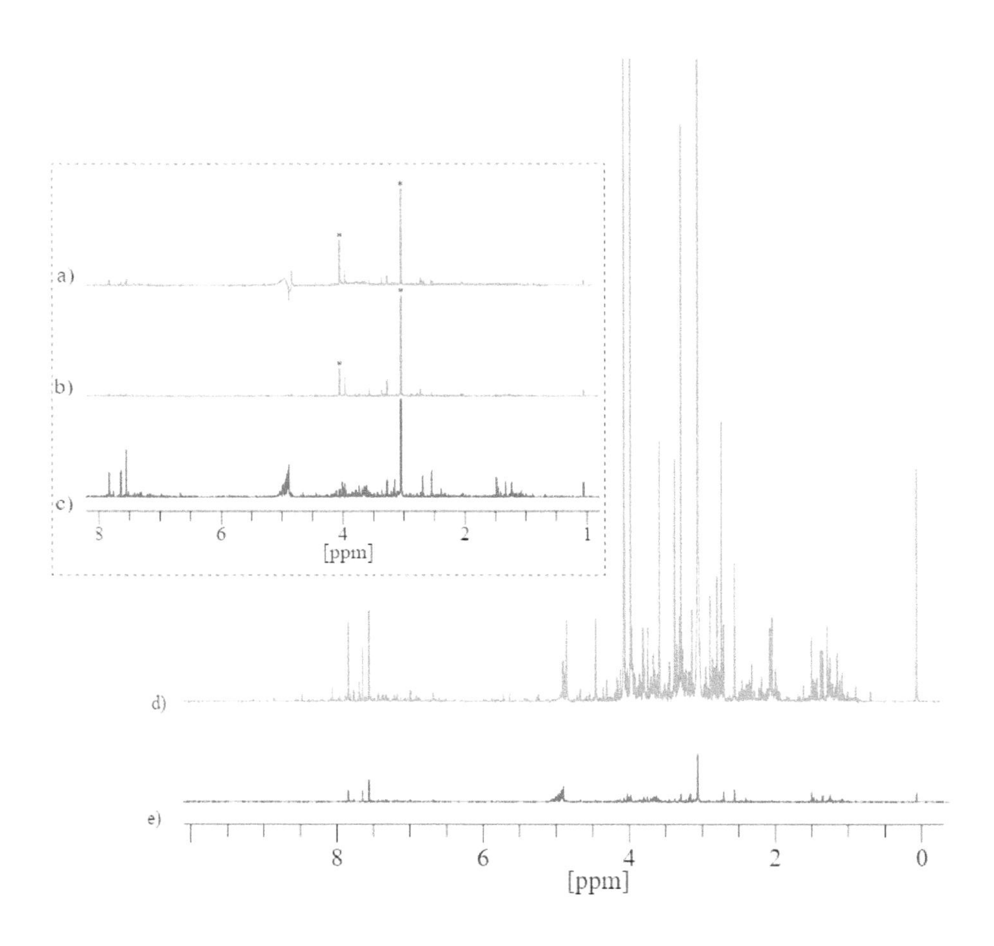

Fig. 3.7. (a)–(c), is a comparison of regular ^{1}H NMR with presaturation, regular *J*-RES projection, and DQF *J*-RES projection respectively from the lyophilized human urine sample in D_2O displayed for the same intensity of

the most intense singlet peak marked with a star. The intense singlet signals marked with star dominate the spectra in (a) and (b) rendering the smaller signals almost undetectable. In contrast, DQF *J*-RES projection in (3.7c) shows great improvement in the intensity of the smaller signals relative to the intense singlet signals whose intensity is now considerably reduced. Comparison on the same noise level of the regular *J*-RES projection and DQF *J*-RES projection is also shown in (d) and (e) which indicates much higher S:N of the regular *J*-RES projection. table 3.3 with the S:N values for five peaks indicates regular *J*-RES has on average six to eight times higher S:N relative to the DQF *J*-RES projection (DQF *J*-RES, in this case, was recorded with the same number of transients as regular *J*-RES). However, the suppression of the intense singlet peaks improves the visibility of the weak intensity metabolite peaks as the dynamic range limitation is much less in the DQF *J*-RES spectrum.

average six to eight times higher S:N relative to the DQF *J*-RES projection. However, the suppression of the intense singlet peaks improves the visibility of the weak intensity metabolite peaks as the receiver gain gets optimised for the smaller signals in DQF *J*-RES spectrum.

Table. 3.3. Lyophilized Human Urine

Note: DQF *J*-RES had equal number of transients in this case than regular *J*-RES

Peaks δ(ppm)	1D proj Reg. JRES (S/N) (A)	1D proj DQF JRES (S/N) (B)	(A)/(B)
1.49	1261	196	6.4
2.39	733	105	6.9
7.54	4647	714	6.5
7.63	2656	414	6.4
6.98	488	57	8.5

4.4 Delay dependent modulation of the signals by the DQF

Due to the presence of a DQ filter at the start of the DQF *J*-RES sequence the intensity of the signals that enter the *J*-RES sequence is modulated by the sinusoidal modulation of the antiphase coherence. Since in a metabolomics complex mixture, large variations exists in *J*-

values due to the presence of a large number of metabolites, thus some signals can disappear in the DQF *J*-RES spectrum when their sine terms approach zero values. Therefore, for certain *J*-values the creation of double quantum coherence is maximum, however, for other *J*-values this could be minimum. In order to address the disappearance of certain multiplet peaks in DQF *J*-RES spectrum, we investigated the delay dependent modulation of the signals by the DQF (using DQF *J*-RES pulse sequence using z-filter as shown in Figure 1. A series of DQF spectra were recorded on the same lyophilized human urine sample and reported in Figure 3.8 using the pulse sequence in Fig. 1 of the main ms but with the ZQ-filter on. Some portions of regular *J*-RES spectrum are displayed in Figures 3.8A, 3.8F, and 3.8K respectively, and their corresponding portions from DQF *J*-RES spectrum in Figures 3.8(B-E), 3.8(G-J), and 3.8(L-O) respectively for four different Δ delays of 20, 30, 40, and 50 ms respectively.

We can draw attention to the dd inside dotted box A_4 (right side) in DQF *J*-RES for Δ=40 ms (Fig 3.8D) - which is absent for Δ=20 and 30 ms (A_2 in Fig 3.8B, A_3 in Fig 3.8C), along with more peaks for Δ=50 ms (A_5 in Fig 3.8E), and not well resolved in regular *J*-RES spectral portion in 3.8A. Similarly for Δ=30 ms the peaks along dotted vertical line B_3 are absent, but better visible along B_2(Δ=20 ms), B_4(Δ=40 ms), and B_5(Δ=50 ms), The peaks along line C_2 and C_3 are very weak but has much better visibility in C_5 and intermediate along C_4 (also distorted along C_1 of regular *J*-RES). Similarly doublet F_2 and F_3 are clear and have good intensity for Δ=20 and 30 ms but loses intensity sharply for Δ=40 (F_4) and 50 ms (F_5), weak in intensity and confusing along F_1 in 3.8F due to the singlets along F_1=0. Finally, dd along E_2 (Δ=20 ms) is evident but unresolved in regular *J*-RES (E_1) and merged with more multiplet lines along E_3, E_4 and E_5.

Overall more multiplet information can be obtained by a combined analysis of these DQF *J*-RES recorded with different values of Δ and comparing with each other as well as to the regular *J*-RES. This process is although time consuming can be fruitful. Some peaks gains intensity while others can loss intensity or even vanish for different values of the delays. But overall the effect of singlet suppression and better visibility of the weak multiplets is maintained for all delays. Overall, along with singlet suppression, delay modulation may improve the results.

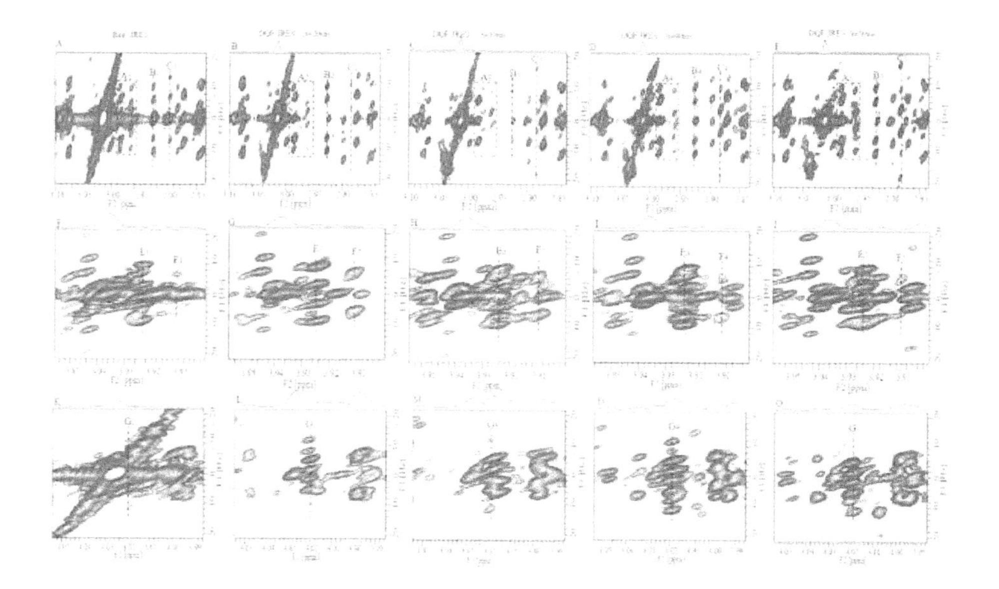

Fig. 3.8. A series of DQF spectra were recorded on the lyophilized human urine sample. Some portions of regular *J*-RES spectrum are displayed in Figures 3.8A, 3.8F, and 3.8K respectively, and their corresponding portions from DQF *J*-RES spectrum are displayed in Figures 3.8(B-E), 3.8(G-J), and 3.8(L-O) respectively for four different Δ in the DQF period of the pulse sequence in Fig 1 for delays of 20, 30, 40, and 50 ms respectively.

4.5 Drawbacks of DQF *J*-RES spectroscopy

4.5.1 G. Pedunculata dried fruit sample

In the result and discussion section of the ms we have shown that DQF *J*-RES method provides better resolution and spectral quality than regular *J*-RES method in G. Pedunculata dried fruit sample in Fig. 3.4. But suppression of multiplets also occurred in addition to suppression of singlet in a few regions of DQF *J*-RES spectrum. These disappeared multiplets in DQF *J*-RES spectrum are indicated by arrows in the portions shown in figures 3.9D, 3.9E, and 3.9F. In the expanded portion of regular *J*-RES spectrum in fig 3.9A, arrows a, b, c, d, and e show the multiplets while the corresponding region of DQF *J*-RES spectrum in Fig 3.9D, these multiplets have disappeared. The same result is observed when we compare expanded regions 3.9B vs 3.9E; and 3.9C vs 3.9F. Thus, a combined analysis using regular *J*-RES and DQF *J*-RES may be richer in information content. While for most of the peaks DQF *J*-RES spectrum is superior, for some peaks the regular *J*-RES spectrum will still have an advantage.

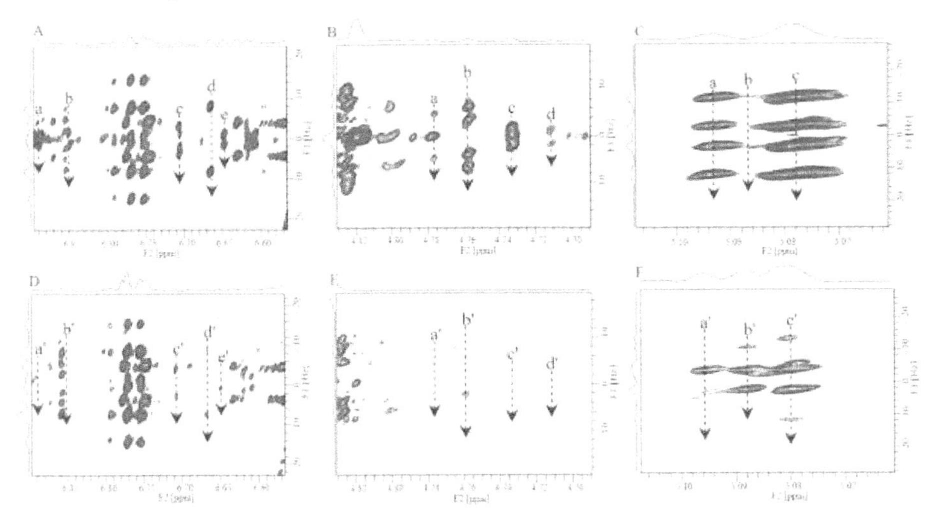

Fig. 3.9. Figures 3.9A, 3.9B and 3.9C are expanded small portion of 45° tilted regular *J*-RES spectrum of G. Pedunculata dried fruit sample in D_2O. The corresponding regions from DQF *J*-RES spectrum are shown in figures 3.9D, 3.9E, and 3.9F respectively. Many peaks marked with small letters in regular *J*-RES spectral portions are absent in the corresponding DQF-*J*-RES spectral portions.

4.5.2 Lyophilized Urine Sample

Suppression of few multiplet peaks also occurred in lyophilized human urine sample in D_2O for the DQF *J*-RES spectrum. These disappeared multiplets in DQF *J*-RES spectrum are indicated by arrows in the portions shown in figures 3.10D, 3.10E, and 3.10F. In the expanded portion of regular *J*-RES spectrum in fig 3.10A, arrows a, b, c, d, and e show the multiplets while the corresponding region of DQF *J*-RES spectrum in Fig S3D, these multiplets have disappeared. The same result is observed when we compare expanded regions 3.10B vs 3.10E; and 3.10C vs 3.10F. Thus, a combined analysis using regular *J*-RES and DQF *J*-RES may be more fruitful. Thus DQF *J*-RES spectrum despite its superior advantages will not completely replace regular *J*-RES spectrum.

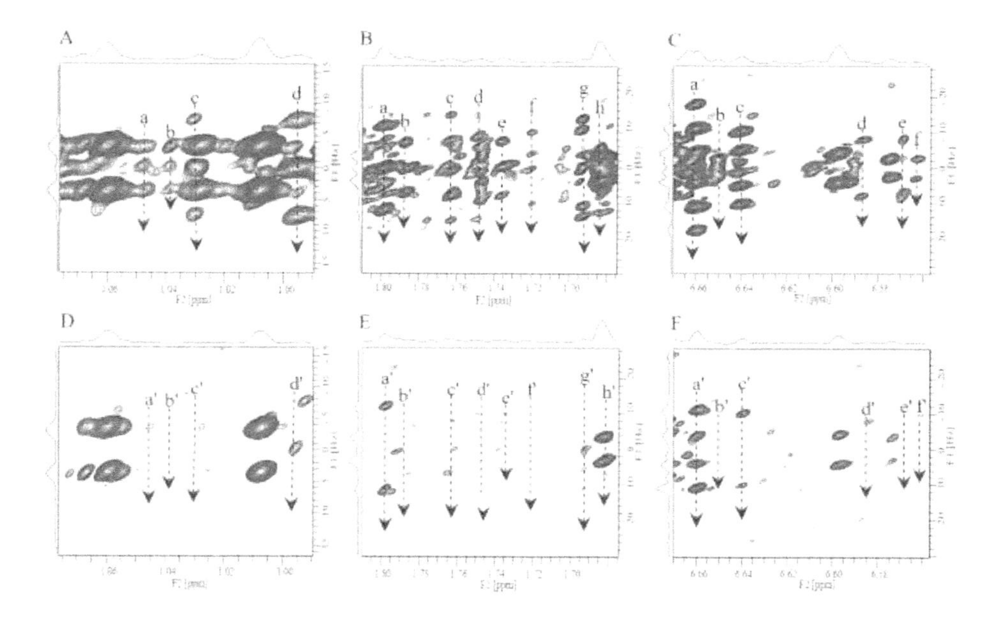

Fig. 3.10. Figures 3.10A, 3.10B and 3.10C are expanded small portion of 45° tilted regular *J*-RES spectrum of lyophilized urine sample in D_2O. The corresponding regions from DQF *J*-RES spectrum are shown in figures 3.10D, 3.10E, and 3.10F respectively. Many peaks marked with small letters in regular *J*-RES spectral portions are absent in the corresponding DQF *J*-RES spectral portions.

3.11A

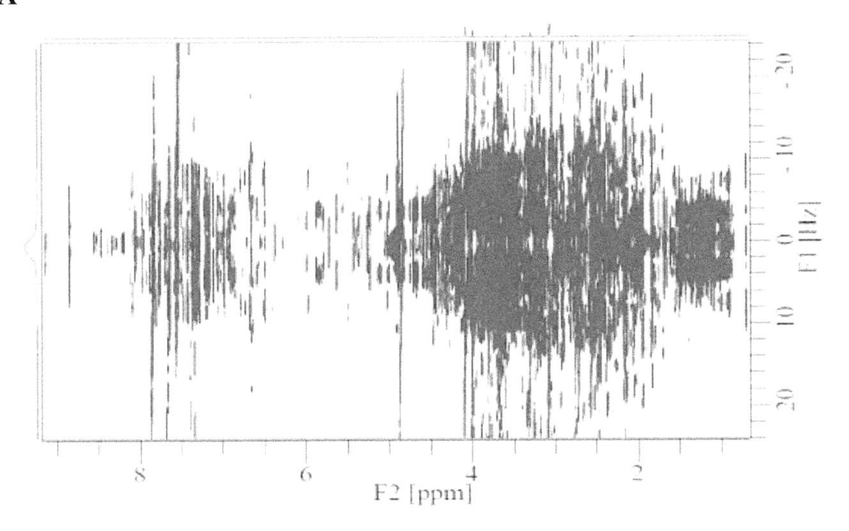

3.11B

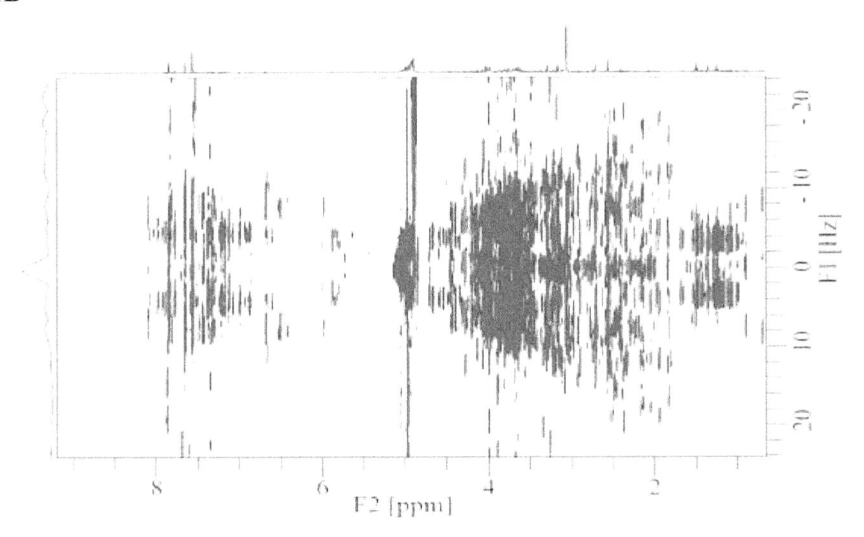

Figure 3.11. A and B are the 45° tilted magnitude mode *J*-RES spectrum and DQF *J*-RES spectrum respectively recorded on lyophilized human urine sample in D_2O. F_1 and F_2 projections are also displayed. Comparison of the F_2 projections indicate that regular *J*-RES spectrum is dominated by a few singlets of high intensity whereas DQF *J*-RES projection shows a large number of signals with good visibility.

3.12A

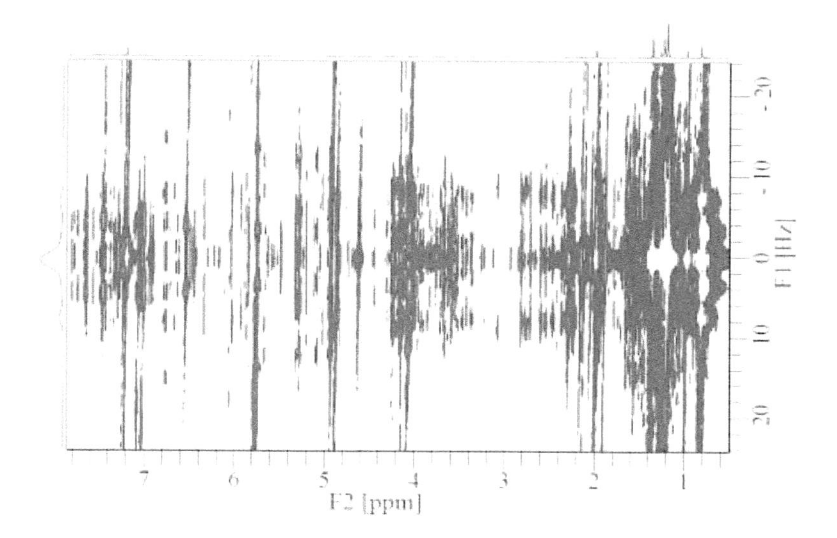

3.12B

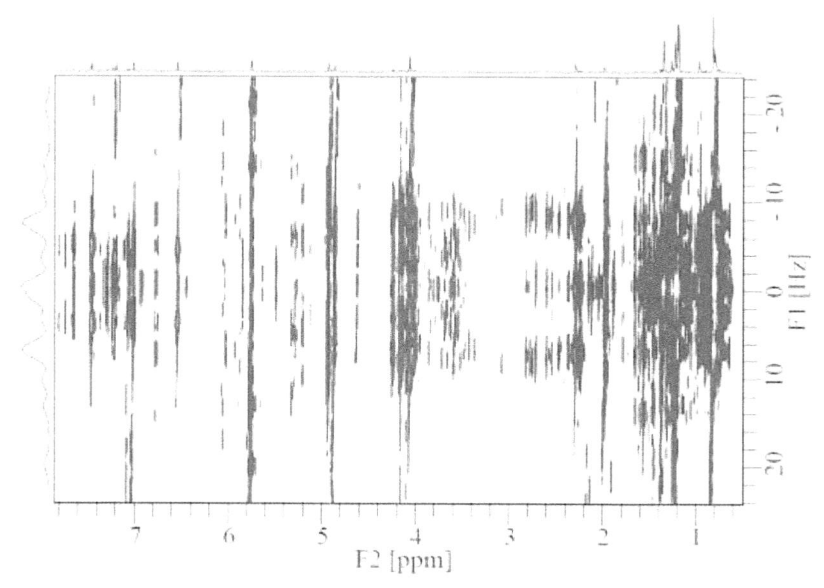

Figure 3.12. A and B are the 45° tilted magnitude mode *J*-RES spectrum and DQF *J*-RES spectrum respectively recorded on G. Pedunculata dried fruit extract dissolved in $CDCl_3$. F_1 and F_2 projections are also displayed.

Comparison of the F_2 projections indicate that a large number of peaks are apparent in DQF *J*-RES projection whereas very few peaks in regular *J*-RES F_2 projection.

4.6 Comparison of the DQF-pJRES and ZQ-filtered DQF-pJRES

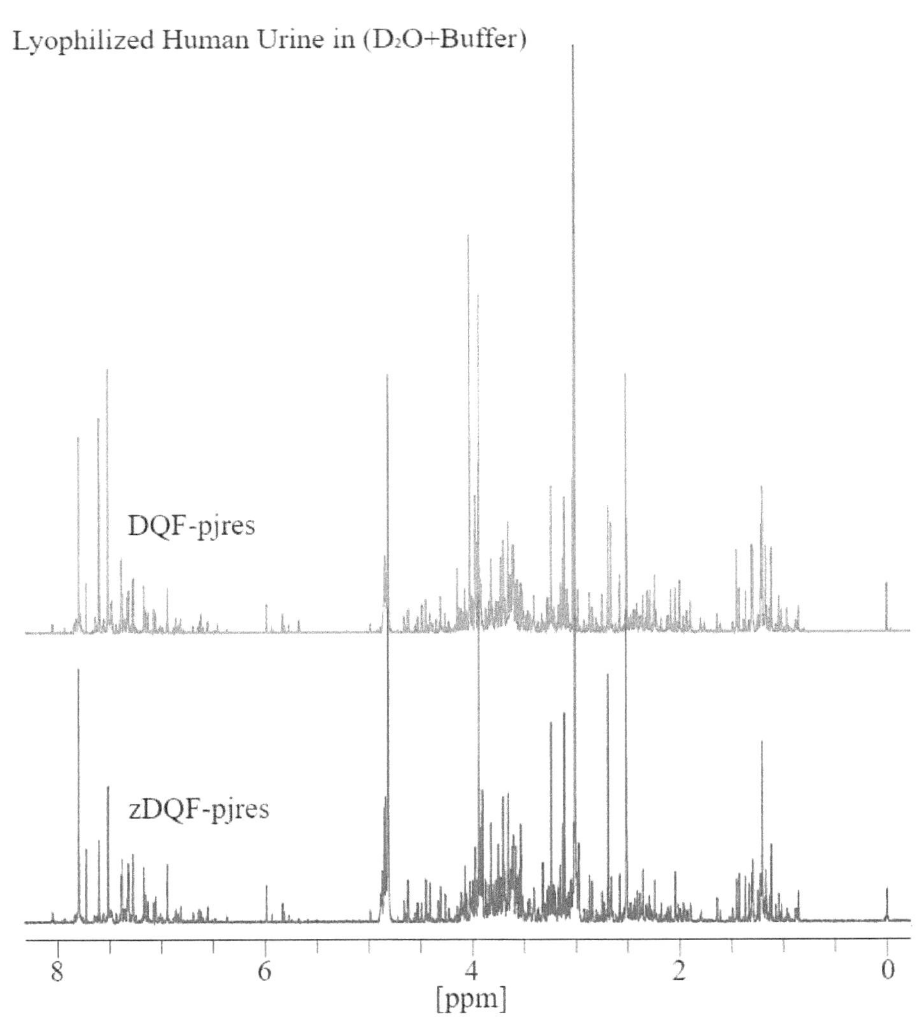

Fig. 3.13. Comparison of the DQF-pJRES (top) and ZQ-filtered DQF-pJRES (bottom) shows small differences of the peak intensities as the ZQ-filter suppress ZQ part of the signals that does not follow the DQ-SQ coherence transfer pathway.

4.7 Pseudo-echo weighting

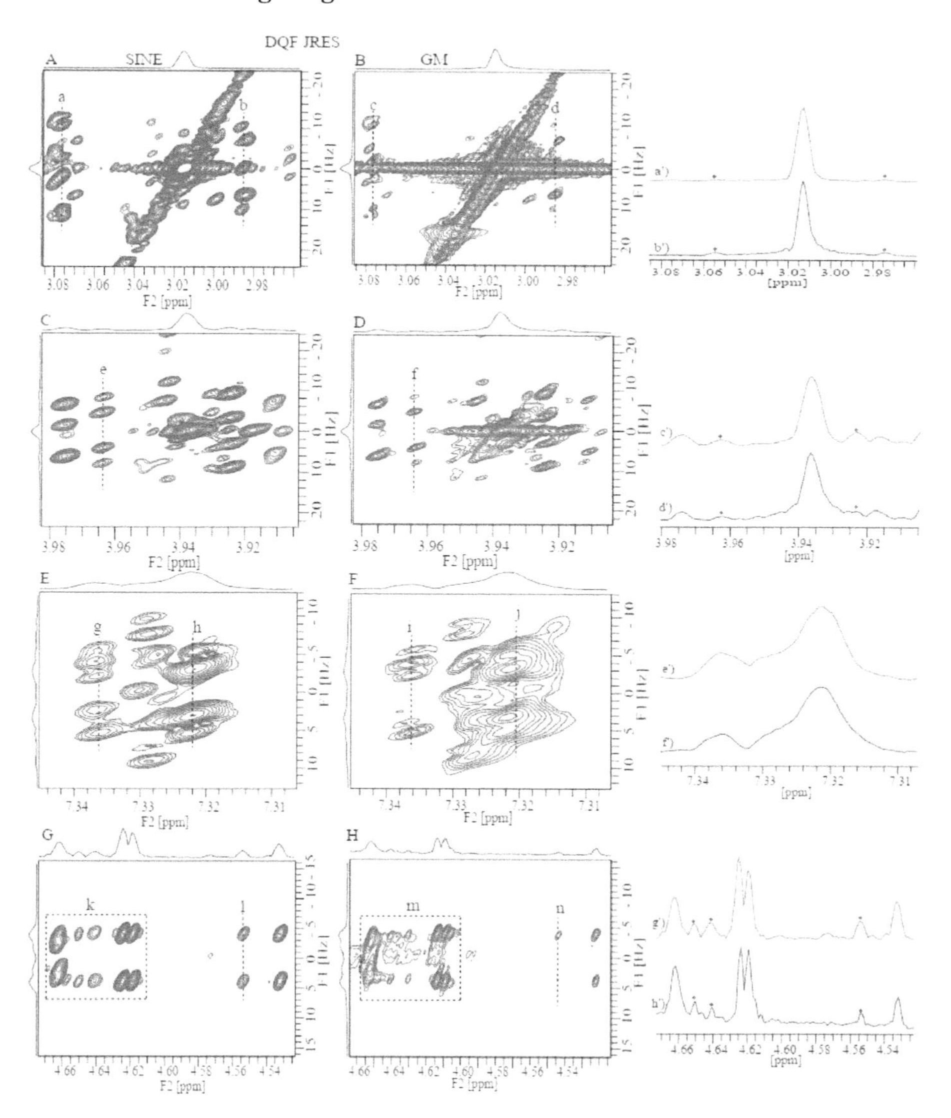

Fig. 3.14. Figures 3.14A, 3.14C, 3.14E and 3.14G are expanded small portion of 45° tilted DQF *J*-RES with z-filter spectrum of lyophilized urine sample in D_2O processed with unshifted sine bell window functions in both F_2 and F_1 dimensions in magnitude mode. The corresponding regions from DQF *J*-RES with z-filter spectrum processed with Gaussian window function (GM) are shown in figures 3.14B, 3.14D, 3.14F, and 3.14H

respectively. Many peaks marked with small letters in middle panels showed minor improvement in resolution. After applying of pseudo-echo window function (GM) in both f_1 and f_2 dimensions prior to Fourier transformation in DQF *J*-RES, distortion of signal 'c' in Fig 3.14B as compare to signal 'a' in Fig 3.14A occurred. Clarity of many signals as 'd', 'f', 'i', 'j' and 'n' are decreased in Fig. 3.14B, 3.14D, 3.14F, and 3.14H respectively as compare to signals 'b', 'e', 'g', 'h' and 'l' in Fig. 3.14A, 3.14C, 3.14E, and 3.14G respectively. Also in corresponding 1D projections on their right hand respectively. A slightly improved resolution in signal 'i' is observed compared to signal 'g' but less improvement in signal 'j' compared to signal 'h' due to their different T_2 relaxation time. The process being iterative separate optimization is necessary for each peaks.

5 Conclusion

We have demonstrated a novel pulse technique the DQF *J*-RES spectroscopy which overcomes the limitations of regular *J*-RES experiments when applied to complex mixtures, where large singlet components of a few signals obscure neighboring important multiplet peaks and also lowers their intensity due to the dynamic range issues. The new technique could clearly identify many important multiplet peaks from sparsely present metabolites in various complex mixtures such as urine and plant extract samples. The new method not only suppressed the intense singlet peaks but also improved the visibility of the weaker multiplet peaks as the receiver gain could be raised in the absence of the strong singlet signals such as creatinine and other peaks in urine. When com- pared for the same noise level, the DQF *J*-RES, in general, displayed six to eight fold reduced S:N, however, the signals are much larger than the noise due to reduced digitization noise. This combined with the significantly reduced intensity of the strong singlet peaks improve the spectral readout of the weaker multiplets in DQF *J*-RES spectra. While most of the signals in each complex mixture studied were better resolved in the DQF *J*-RES spectrum; however, a few peaks were absent, which could be detected in regular *J*-RES spectrum. This effect was also systematically studied for various values of the DQF period and revealed more information on the multiplets could be acquired by recording a series of DQF *J*-RES spectra with a few different delays for the DQ filter period. Thus a combined analysis using regular *J*-RES and DQF *J*-RES with different delays may be more fruitful.

6. Acknowledgments

Upendra Singh thanks UGC (University Grants Commission) India, for a research fellowship. We thank the Science and Engineering Research Board (SERB) under the Department of Science & Technology (DST), Govt. Of India, DST No: EMR/2014/001280 (Grant No. SERB/F/6435/2015-16) for the extramural research grant.

7. Future Work

This method can be utilized in identification of metabolites in the complex mixtures such as plant extracts and bio-fluids as plasma, serum, urine.

References:

[1] J. M. Fonville, A.D. Maher, M. Coen, E. Holmes, J.C. Lindon, J.K. Nicholson, Evaluation of full-resolution J-resolved 1H NMR projections of biofluids for metabonomics information retrieval and biomarker identification, Anal. Chem. 82 (2010) 1811–1821.

[2] P.J.D. Foxall, J.A. Parkinson, I.H. Sadler, J.C. Lindon, J.K. Nicholson, Analysis of biological fluids using 600 MHz proton NMR spectroscopy: application of homonuclear two-dimensional *J*-resolved spectroscopy to urine and blood plasma for spectral simplification and assignment, J. Pharm. Biomed. Anal. 11 (1993) 21–31.

[3] P.J.D. Foxall, M. Spraul, R.D. Farrant, L.C. Lindon, G.H. Neild, J.K. Nicholson, 750 MHz 1H-NMR spectroscopy of human blood plasma, J. Pharm. Biomed. Anal. 11 (1993) 267–276.

[4] B.C. Sweatman, R.D. Farrant, E. Holmes, F.Y. Ghauri, J.K. Nicholson, J.C. Lindon, 600 MHz 1H-NMR spectroscopy of human cerebrospinal fluid: Effects of sample manipulation and assignment of resonances, J. Pharm. Biomed. Anal. 11 (1993) 651–664.

[5] M.J. Lynch, J. Masters, J.P. Pryor, J.C. Lindon, M. Spraul, P.J.D. Foxall, J.K. Nicholson, Ultra high field NMR spectroscopic studies on human seminal fluid, seminal vesicle and prostatic secretions, J. Pharm. Biomed. Anal. 12 (1994) 5– 19.

[6] H. Antti, T.M.D. Ebbels, H.C. Keun, M.E. Bollard, O. Beckonert, J.C. Lindon, J.K. Nicholson, E. Holmes, Statistical experimental design and partial least squares regression analysis of biofluid metabonomic NMR and clinical chemistry data for screening of adverse drug effects Chemom, Intell. Lab. Syst. 73 (2004) 139– 149.

[7] M.R. Viant, C. Ludwig, S. Rhodes, U.L. Guenther, D. Allaway, Validation of a urine metabolome fingerprint in dog for phenotypic classification, Metabolomics 3 (2007) 453–463.

[8] M.R. Viant, Improved methods for the acquisition and interpretation of NMR metabolomic dat, Biochem. Biophys. Res. Commun. 310 (2003) 943–948.

[9] J.K. Nicholson, P.J.D. Foxall, M. Spraul, R.D. Farrant, J.C. Lindon, 750 MHz 1H and $^{1}H–^{13}C$ NMR spectroscopy of human blood plasma, Anal. Chem. 67 (1995) 793–811.

[10] H.T. Widarto, E. Van der Meijden, A.W.M. Lefeber, C. Erkelens, K. Kim, H.Y.H. Choi, R. Verpoorte, Metabolomic differentiation of Brassica rapa following herbivory by different insect instars using two-dimensional nuclear magnetic resonance spectroscopy, J. Chem. Ecol. 32 (2006) 2417–2428.

[11] M.R. Viant, J.G. Bundy, C.A. Pincetich, J.S. de Ropp, R.S. Tjeerdema, NMR- derived developmental metabolic trajectories: an approach forvisualizing the toxic actions of trichloroethylene during embryogenesis, Metabolomics 1 (2005) 149–158.

[12] A. Khatib, E.G. Wilson, H.K. Kim, A.W.M. Lefeber, C. Erkelens, Y.H. Choi, R. Verpoorte, Application of two-dimensional *J*-resolved nuclear magnetic resonance spectroscopy to differentiation of beer, Anal. Chim. Acta 559 (2006) 264–270.

[13] C. H. Johnson, T.J. Athersuch, I.D. Wilson, L. Iddon, X.L. Meng, A.V. Stachulski, J. C. Lindon, J.K. Nicholson, kinetic and *J*-resolved statistical total correlation NMR spectroscopy approaches to structural information recovery in complex reacting mixtures: application to acyl glucuronide intramolecular transacylation reactions, Anal. Chem. 80 (2008) 4886–4895.

[14] Y.L. Wang, M.E. Bollard, H. Keun, H. Antti, O. Beckonert, T.M. Ebbels, J.C. Lindon, E. Holmes, H.R. Tang, J.K. Nicholson, Spectral editing and pattern recognition methods applied to high-resolution magic-angle spinning 1H nuclear magnetic resonance spectroscopy of liver tissues, Anal. Biochem. 323 (2003) 26–32.

[15] J. Farjon, D. Merlet, P. Lesot, J. Courtieu, Enantiomeric excess measurements in weakly oriented chiral liquid crystal solvents through 2D 1H selective refocusing experiments, J. Magn. Reson. 158 (2002) 169–172.

[16] K. Kobzar, H. Kessler, B. Luy, Stretched gelatin gels as chiral alignment media capable to discriminate enantiomers, Angew. Chem. Int. Ed. 44 (2005) 3145– 3147. Corrigendum: Angew, Chem. Int. Ed. 44 (2005) 3509.

[17] S. R. Chaudhari, N. Suryaprakash, Chiral discrimination and the measurement of enantiomeric excess from a severely overcrowded NMR spectrum, Chem. Phys. Lett. 555 (2013) 286–290.

[18] B. Luy, J.P. Marino, JE-TROSY: comined *J*- and TROSY-spectroscopy for the measurement of one-bond couplings in macromolecules, J. Magn. Reson. 163 (2003) 92–98.

[19] J. Furrer, M. John, H. Kessler, B. Luy, *J*-spectroscopy in the presence of residual dipolar couplings: determination of one-bond coupling constants and separate dimension with scalable resolution, J. Biomol. NMR 37 (2007) 231–243.

[20] M. Nilsson, G.A. Morris, Pure shift proton DOSY: diffusion-ordered ^{1}H spectra without multiplet structure, Chem. Commun. 9 (2007) 933–935.

[21] A.J. Pell, J. Keeler, Two-dimensional J-spectra with absorption-mode lineshapes, J. Magn. Reson. 189 (2007) 293–299.

[22] S. Simova, H. Sengstschmid, R. Freeman, Proton chemical-shift spectra, J. Magn. Reson. 124 (1997) 104–121.

[23] A. Bax, A.F. Mehlkopf, J. Smidt, A fast method for obtaining 2D *J*-resolved absorption spectra, J. Magn. Reson. 40 (1980) 213–219.

[24] V.A. Mandelshtam, Q.N. Van, A.J. Shaka, Obtaining proton chemical shifts and multiplets from several 1D NMR signals, J. Am. Chem. Soc. 120 (1998) 12161– 12162.

[25] B. Luy, Adiabatic z-filtered *J*-spectroscopy for absorptive homonuclear decoupled spectra, J. Magn. Reson. 201 (2009) 18–24.

[26] A. Verma, B. Baishya, Real-time band-selective homonuclear proton decoupling for improving sensitivity and resolution in phase-sensitive *J*-resolved spectroscopy, ChemPhysChem. 16 (2015) 2687–2691.

[27] M. Foroozandeh, R.W. Adams, P. Kiraly, M. Nilsson, G.A. Morris, Measuring couplings in crowded NMR spectra: pure shift NMR with multiplet analysis, Chem. Commun. (2015) 1–3.

[28] A. Verma, B. Baishya, Real-time bilinear rotation decoupling in absorptive mode *J*-spectroscopy: detecting low-intensity metabolite peak close to high-intensity metabolite peak with convenience, J. Magn. Reson. 266 (2016) 51–58.

[29] S. H. Smallcombe, S.L. Patt, P.A. Keifer, WET solvent suppression and its applications to LC NMR and high-resolution NMR spectroscopy, J. Magn. Reson. Series A 117 (1995) 295–303.

[30] M. Rance, O.W. Sørensen, G. Bodenhausen, G. Wagner, R.R. Ernst, K. Wüthrich, Improved spectral resolution in COSY ^{1}H NMR spectra of proteins via double quantum filtering, Biochem. Biophys. Res. Commun. 117 (1983) 479–485.

[31] U. Piantini, O.W. Sørensen, R.R. Ernst, Multiple quantum filters for elucidating NMR coupling networks, J. Am. Chem. Soc. 104 (1982) 6800–6801.

[32] A. J. Shaka, R. Freeman, Simplification of NMR spectra by filtration through multiple-quantum coherence, J. Magn. Reson. 51 (1983) 169–173.

[33] N. Müller, R.R. Ernst, K. Wüthrich, Multiple-quantum-filtered two-dimensional correlated NMR spectroscopy of proteins, J. Am. Chem. Soc. 108 (1986) 6482– 6492.

[34] P.C.M. van Zijl, M. O'Neil Johnson, S. Mori, R.E. Hurd, Magic-angle-gradient double-quantum-filtered COSY, J. Magn. Reson., Ser A 113 (1995) 265–270.

[35] A. L. Davis, E.D. Laue, J. Keeler, D. Moskau, J.J. Lohman, Absorption-mode two dimensional NMR spectra recorded using pulsed field gradients, J. Magn. Reson. 94 (1991) 637–644.

[36] Jacob M. Newman, Alexej Jerschow, Improvements in complex mixture analysis by NMR: DQF-COSY iDOSY, Anal. Chem. 79 (2007) 2957–2960.

[37] Y. Huang, S. Cai, Z. Zhang, Z. Chen, High-resolution two-dimensional *J*-resolved NMR spectroscopy for biological systems, Biophys. J. 106 (2014) 2061–2070.

[38] Y. Huang, Z. Zhang, H. Chen, J. Feng, S. Cai, Z. Chen, A high-resolution 2D *J*-resolved NMR detection technique for metabolite analyses of biological samples, Sci. Reports 5 (2015) 8390.

Chapter 4

Pure shift HMQC: Resolution and sensitivity enhancement by bilinear rotation decoupling in the indirect and direct dimensions

1. Introduction

Homonuclear scalar coupling provides crucial structural information. However, its presence degrades the sensitivity and resolution of proton 1D as well as the 2D spectrum. Lack of resolution has driven the rapid growth of pure shift (ps)[1–34] NMR spectroscopy in recent years. Consequently, when coupling information is not being sought, ps ^{1}H spectrum is desirable in both 1D as well as 2D homonuclear as well as heteronuclear correlation experiments that involve ^{1}H either from a need to attain ultrahigh- resolution or/and enhanced sensitivity.

The broadband pure shift approaches have found significant attention in recent years and comprise of three schemes-(1) Zangger-Sterk (ZS) method[1–10] reduces the ^{1}H multiplets to singlets by relying on spatially encoded spectrally selective inversion of a fraction (1–10%) of ^{1}H magnetization aided by field gradients. (2) BIRD[11–24] based broadband ^{1}H-^{1}H homo-decoupling utilizes the 1.1% ^{1}H signals attached to natural abundant ^{13}C nuclei for obtaining the ps spectrum. Suitable BIRD blocks combined with a non-selective ^{1}H hard refocusing pulse delivers broadband homo- decoupled ^{1}H ps spectrum. (3) The PSYCHE (Pure Shift Yielded by Chirp Excitation)[29–34] scheme also refocuses a fraction of magnetization (active spins) using a stimulated echo via two consecutively applied low flip angle, swept-frequency chirp pulses in the presence of field gradient. In this process rest of the magnetization (passive spins) are left unperturbed. Chemical shift evolution can be executed when this process is immediately followed by a hard 180° pulse while retaining the broadband homonuclear refocusing effect. The fraction of active spins are often bigger than those of ZS and BIRD and depends on the flip angles used (3–20%). However, sensitivity can be equal to those of ZS when the cleaner spectrum is desired, as the flip angle needs to be optimized to a small value for achieving a cleaner spectrum. There are also band selective methods such as BASHD and HOBS[25–28], which produce the ps spectrum for a selected signal or a group of signals in the selected region. Although they offer the highest sensitivity, they are not broadband.

Various two-dimensional correlation experiments, such as TOCSY[6,30,20], NOESY[7,32], COSY[7,34], and HSQC[14,15,16,18,44] have been upgraded for high resolution using ps techniques such as ZS, BIRD, and PSYCHE. Pure shift NMR has also been utilized for recording absorptive mode *J*-resolved spectroscopy for accessing highly resolved *J*-coupling information in crowded spectral regions, which otherwise gets obscured due to dispersive contribution in regular *J*-resolved spectroscopy[35–43,19].

There are also other approaches, such as constant time[58], *J*-Resolved NMR[60], time-reversal[59], perfect-echo[16,49–57], and SITCOM effect[45–48] for suppressing the effect of homonuclear scalar couplings.

Data collection for the ps spectrum is carried out generally using either a interferogram approach or a real-time approach. In the interferogram approach, a series of FIDs are recorded in a pseudo-2D fashion. In general, the homonuclear ^1H-^1H scalar cou- pling is small. For a 10 Hz coupling, less than 5% signal modulation occurs for a data chunk of duration 10 ms. Consequently, 30 to 40 data chunks of 10 to 20 ms duration (with carefully chosen chemical shift evolution and *J*-refocusing) collected in a pseudo-2D mode can be concatenated to form a synthetic FID which when Fourier Transformed provides ^1H ps spectrum. In the real-time mode (or single scan mode), data acquisition is interrupted after every 10 to 20 ms duration in a single FID so that scalar coupling evolution is negligible throughout the FID while chemical shift evolution is continuous between the non-interrupted periods. The pure shift spectrum can also be obtained from the F_1 projection of the F_1-decoupled spectrum. BIRD, ZS, and HOBS can be acquired in real-time mode as well as in interferogram mode. PSYCHE is applicable in interferogram mode only.

Application of broadband ps NMR is limited by the low sensitivity inherent to the techniques. In some cases, artifact suppression and sensitivity are inversely related. Until now, real-time BIRD- HSQC is the only approach that does not pay any sensitivity penalty while acquiring broadband homo-decoupled ^1H dimension as it already relies on 1.1% ^1H signals attached to ^{13}C at natural abundance (or ^{15}N labeled proteins as well).

Real-time BIRD-HSQC has found manyapplication due to this aspect. An important heteronuclear correlation experiment in NMR is the HMQC experiment[61–62]. HMQC is still routinely used for small molecules, as it is less sensitive to pulse miscalibration than HSQC. Numerous applications exist for studying proteins, nucleic acids, and other macromolecules as the experiment is suit- able for small flip angle and selective pulses such as so-fast HMQC[63] and small flip angle HMQC[64]. Another advantage of ^1H-^{15}N HMQC in labeled proteins is the slower relaxation of heteronuclear multiple quantum coherence relative to the transverse relaxation time of amide protons and had led to the development of experi- ments for measuring NH-CaH J-coupling including Constant-Time HMQC variants[65–69]. When a high resolution is being sought along F_1 as well as F_2 dimensions, HSQC is more preferred due to higher sensitivity. This is because HSQC has ^1H-^1H multiplet structures only along F_2, whereas HMQC has multiplet structures along F_1 as well as F_2.

In the present work, we focus on improving the resolution and sensitivity of the ^1H-^{13}C HMQC experiment using BIRD pulses. Generally, BIRD is not applicable at the center of t_1-evolution in HMQC as multiple quantum is not modulated by (^1JCH) direct heteronu- clear coupling, which is a prerequisite of the BIRD element. How- ever, an initial t_1 period can be added to ^1H single quantum evolution before multiple quantum evolution in the 2nd t_1 period. A BIRD block separates the two t_1 periods. This way, one can enhance sensitivity and resolution along the F_1 dimension of the HMQC experiment when the multiplet structure is well resolved in F_1. Further, real-time BIRD based pure shift acquisition during t_2 of HMQC (similar to the standard real-time BIRD HSQC) should improve its resolution and sensitivity irrespective of short or long t_1 acquisition times.

2. Pulse sequence of ps-HMQC

Fig. 4.1A displays the pulse sequence for obtaining the ps-HMQC spectrum. The conventional HMQC pulse sequence is also shown in 4.1B. Both experiments were recorded in the phase-sensitive mode using the echo-antiecho approach for frequency discrimination utilizing field gradients[70–74]. In the ps-HMQC pulse sequence, there are two t_1 evolution period- first one between time points b and c, and the 2nd one between the time points g and h. ^{13}C chemical shift evolution takes place in the 2nd t_1 period between time points g and h. Homonuclear ^1H-^1H scalar coupling evolves between time points a and d ($=t_1 + 2\Delta + 2\Delta_3$) which gets refocused between time points e and j ($=t_1 + 2\Delta + 2\Delta_3$) due to the application of the BIRD block between time points d and e. Thus there is an extra proton SQ coherence evolution period present in this sequence, which is generally absent in the regular HMQC shown in Fig. 4.1B. A product operator calculation is given in the supporting information. In regular HMQC (4.1B), ^{13}C–^1H ^1J-coupling evolves between time points e and f ($=\Delta$ tuned to $^1J_{CH}$), and heteronuclear multiple quantum coherence is created at time point f after the 1st $90°$ ^{13}C pulse at time point f. ^{13}C chemical shift evolves from time point g to h, while the ^1H

chemical shift is refocused by the refocusing pulse at the center of the t_1 period between time point g and h. Further, this refocusing pulse also refocuses the ^{13}C–^{1}H ^{n}J-couplings (where n > 1), and ^{13}C–^{1}H multiple-quantum coherence does not evolve under ^{13}C–^{1}H ^{1}J-couplings (hence BIRD does not apply to multiple quantum coherence). However, ^{1}H magnetization being transverse evolves under ^{1}H-^{1}H J-couplings between time points e to j. Similar coherence evolution occurs in the ps-HMQC sequence (in 4.1A) also from time point e to j, but in this case refocusing of ^{1}H-^{1}H J-couplings takes place between time points e and j, instead of evolution, due to the application of BIRD element and the prefocusing of J_{HH} from a to d. In regular HMQC, ^{1}H-^{1}H J-couplings lead to the creation of homonuclear antiphase terms (such as $I_{1Y}I_{2z}$) with dispersive line shape,

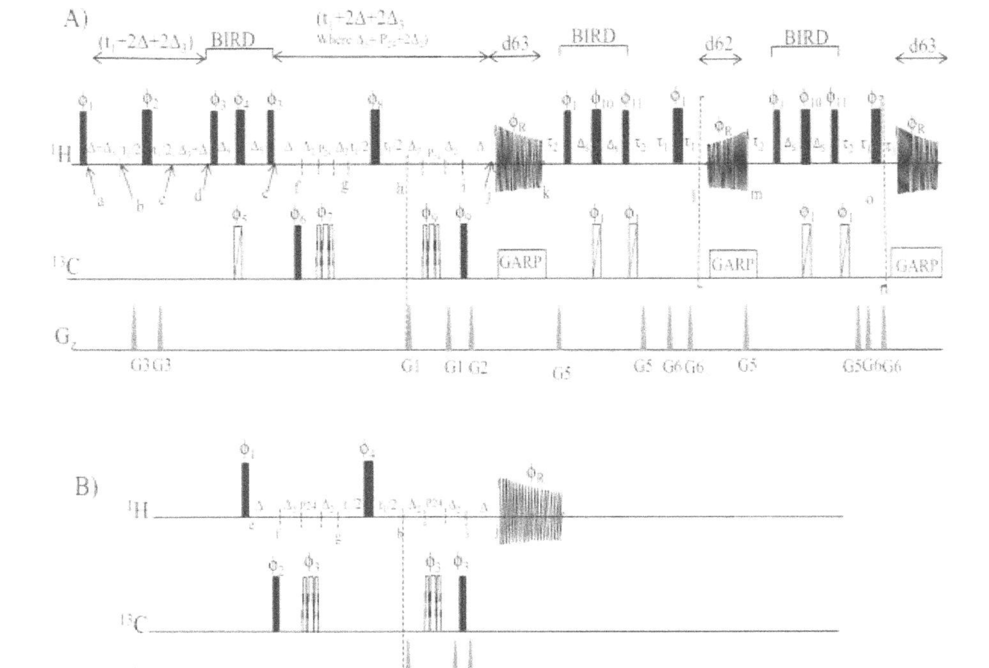

Fig. 4.1. (A) ps-HMQC pulse sequence with BIRD elements. In the ps-HMQC, a BIRD element divides the HMQC period ($t_1 + 2\Delta + 2\Delta_3$) and a spin echo period equal to the HMQC period. A BIRD element replaces the 1st 90 pulse of regular HMQC. Narrow and wide filled rectangular black bars are hard 90° and 180° pulses. The ^{13}C inversion pulses during all BIRD elements are Bip.720,100,10.1 pulse (of duration 240 ls each) and the $_{13}C$ refocusing pulses during the two delays p24 are a pair of smoothed chirp composite pulses (60 kHz total sweep

width, 2.0 ms length, four chirp composite pulse, Crp60comp.4) which together acts as a phase-compensating adiabatic pulse pair. Gradients with SMSQ10.100 shape are G_1 = 80% for 1 ms, G_2 = 40.2% for 1 ms, G_3 = 17% for 1 ms G_5 = 3% for 300 ms and G_6 = 5% for 300 ms. Delay $\Delta = 1/(2 * {}^1J^{13}{}_{C\text{-}H})$, $\Delta_5 = 2\Delta$ and $\Delta_3 = P24 + 2\Delta_2$ (where Δ_2 is the duration of the gradient G_1 and the corresponding gradient recovery delay). τ_2 and τ_1 are small delays to accommodate the G_5 and G_6 gradients. GARP is used for decoupling during acquisition. Echo-Antiecho for Frequency discrimination in t_1 using gradients G_1 and G_2. Windowed data acquisition is performed during t_2. The FID is interrupted periodically by the application of BIRD elements, which refocuses the effects of ^{1}H-^{1}H *J*-couplings. Phase cycling: ϕ_1 = x, ϕ_2 = x, ϕ_3 = x, ϕ_4 = y, ϕ_5 = x, ϕ_6 = x -x, ϕ_7 = x, ϕ_8 = x, ϕ_9 = x x -x -x, ϕ_{10} = y, ϕ_{11} = -x, ϕ_{30} = x, and ϕ_R = x -x -x x. Product operator calculation and various acquisition and processing parameters of the pulse sequence along with the pulse program are given in Table 4.1 and table 4.2 for 1, 8-dihydroxy-3-methylanthracene-9,10-dione and glucose molecules respectively. (B) Conventional HMQC pulse sequence with time points matched to the ps-HMQC sequence. ϕ_1 = x, ϕ_2 = x -x, ϕ_3 = 4(x) 4(-x), ϕ_4 = x x-x -x , ϕ_R = x -x x -x -x x -x x. Echo-Antiecho for Frequency discrimination in t_1 using gradients G_1 and G_2 as in A.

which along with in-phase absorptive part creates problem in attaining a purely absorptive 2D HMQC spectrum. Generally, these dispersive terms are eliminated by a 90°y purge pulse just before the acquisition or by a killer gradient in a modified sequence. However, this process leads to loss of useful signal and hence, sensitivity. The BIRD decoupling during t_1 in ps-HMQC refocuses these antiphase dispersive terms and enhance the sensitivity of the experiment along with multiplet to singlet conversion. The real-time application of the BIRD block for broadband homo- decoupling during the t_2 period is straightforward and are detailed in earlier literature[17,18,75], which further enhances sensitivity and resolution along F_2 dimension.

CT-HMQC experiment in which multiple quantum coherences are spin locked for signal enhancement are also reported[65]. However, constant-time experiments suffer from signal loss as the cosine modulated signals cannot be maximized for any duration of the constant time period as ^{1}H-^{1}H *J*-couplings vary significantly. These can result in the absence of many cross-peaks for complex organic molecules.

For comparison, we also performed the real-time ps-HSQC experiment that implements real-time pure shift acquisition using BIRD for enhancement of resolution and sensitivity, and the detail pulse sequence of ps-HSQC can be found in recent publications[17,18,75]. Unlike the HMQC spectrum, the HSQC spectrum does not show any multiplet structure along F_1 as the heteronuclear antiphase coherence (single quantum coherence such as $^{13}C_xH_z$) in t_1 dimension of the HSQC does not evolve under ^{1}H-^{1}H *J*-couplings. However, the direct dimension does show splitting due to the ^{1}H-^{1}H *J*-couplings lowering the resolution and sensitivity of the technique. Only recently, with the emergence of real-time pure shift acquisition, this limitation of the HSQC experiment could be overcome, which inspired us to explore similar advantages in the HMQC experiment. Therefore, ps-HSQC was also performed along with the HMQC experiments in Fig. 4.1A and B.

In the implementation of the ps-HMQC and ps-HSQC experiments, there is a challenge associated with the BIRD train applied during the real-time as it involves repeated inversion of the ^{13}C resonances over a spectral window of 200 ppm. In addition, there are two refocusing pulses on ^{13}C during the t_1 dimension of ps-HMQC. The pulse imperfections associated with hard ^{13}C inversion and refocusing pulses can be significant in high fields as there are too many such pulses in the ps-HMQC and ps-HSQC. Therefore, we replaced all the ^{13}C inversion and refocusing pulses by their suit- able broadband versions in order to minimize the artifacts from pulse imperfections analogous to the ps-HSQC experiment[18,75] and real-time BIRD sequence for 1D ps-^{1}H NMR[17]. We replaced the two ^{13}C refocusing pulses during the delays p24 in Fig. 4.1A and 4.1B in the t_1 dimension of HMQC and ps-HMQC by smoothed chirp composite pulses (60 kHz total sweep width, 2.0 ms length, four chirp composite pulse, Crp60comp.4) which together acts as a phase-compensating adiabatic pulse pair. A sin- gle chirp pulse can introduce a quadratic phase to the involved coherence, which can be eliminated if a pair of such chirp is implemented[78,17]. The ^{13}C inversion pulses during all BIRD elements in the ps-HMQC are Broadband Inversion Pulses (BIP 720.100.10.1)[76]. Smoothed linear Chirp (60 kHz total sweep width, 0.5 ms length, 20% smoothing, sweep from high to low field, Crp60,0.5,20.1) pairs are also used during BIRD in some earlier work[17]. The design of broadband inversion and refocusing pulses using composite pulses and adiabatic pulses or a hybrid of the two is well documented in NMR[76–79]. The initial form of BIP pulses is generated from either a linear frequency sweep or a 90y-180x-90y composite pulse template. While the amplitude and duration are kept constant for a given optimization, the phase modulation is numerically optimized to generate a group of rectangular phase-modulated pulses with dual compensation[76]. BIPs have nonlinear frequency sweep and more convenient to use due to a relatively short duration. Further, due to the absence of amplitude modulation linearity of the rf amplifier is not an issue. If the offset range is not huge, BIPs can be used for short and highly efficient inversion. However, for a very large offset range, adiabatic pulses are better, and BIPs tend towards adiabatic pulses but allows nondiabetic trajectories. Adiabatic pulses can perform a wide band of inversion with moderate rf power and also immune to B_0 (static field) and B_1 (rf) inhomogeneities.

3. Product Operators Involved In Pure Shift HMQC Pulse Sequence Shown in Fig. 4.1

Since homonuclear and heteronuclear terms commute, we consider initially the homonuclear evolution during the pulse sequence from time point 'a' to 'j'. Then we will consider the heteronuclear evolution from time point 'e' to 'j'

The operator at time point 'a' is $-I_{1y}$.

Homonuclear coupling evolution takes place between time points 'a' to 'd'

The operators at time point 'd' are

$$-I_{1y}\cos\{\pi jt\} + 2I_{1x}I_{2z}\sin\{\pi jt\} \quad (1)$$

Where t is equal to $t_1 + 2\Delta + 2\Delta_3$, and j is the ^1H-^1H scalar coupling

The BIRD element between time point 'd' to 'e' reverses the sign of the 2nd term as the 12C attached proton is selectively inverted resulting in

$$-I_{1y}\cos\{\pi jt\}-2I_{1x}I_{2z}\sin\{\pi jt\} \quad (2)$$

Which subsequently evolves to the following terms at time point 'j'

$$\text{Cos}(\pi jt)[-I_{1y}\cos(\pi jt)+2I_{1x}I_{2z}\sin(\pi jt)]-\sin(\pi jt)[2I_{1x}I_{2z}\cos(\pi jt)+I_{1y}\sin(\pi jt)] \quad (3)$$

Which can be written as

$$-I_{1y}\cos^2(\pi jt)+2I_{1x}I_{2z}\cos(\pi jt)\sin(\pi jt)-2I_{1x}I_{2z}\sin(\pi jt)\cos(\pi jt)-I_{1y}\sin^2(\pi jt). \quad (4)$$

Second term and third term are equal and cancel to each other due to opposite sign and the I_{1y} terms can be written as

$$-I_{1y}[\cos^2\pi jt+\sin^2\pi jt] = -I_{1y} \quad \text{at the at time point 'j'.} \quad (5)$$

Heteronuclear coupling evolution takes place from the time point 'e' to 'f' for duration Δ.

$$-I_{1y}\cos(\pi j_{CH}\Delta)+2I_{1x}S_z\sin(\pi j_{CH}\Delta). \quad (6)$$

Where j_{CH} is the one bond ^1H-^{13}C scalar coupling, and magnetization S represents carbon nucleus. The value of Δ is adjusted to $1/2j_{CH}$, then equation (6) becomes

$$2I_{1x}S_z \quad (7)$$

90° pulse on ^{13}C, lead to multiple quantum coherence as

$$2I_{1x}S_y \tag{8}$$

During 'g' to 'h' chemical shift evolution of ^{13}C takes place.

$$2I_{1x}S_y\cos(\Omega_s t_1) - 2I_{1x}S_x\sin(\Omega_s t_1) \tag{9}$$

Subsequently, from time point 'i' to 'j' conversion to proton single quantum takes place

$$I_{1y}\cos(\Omega_s t_1) \tag{10}$$

The sine part remains as multiple quantum and is not detected. Thus, the final term at the start of FID acquisition is free from any homonuclear *J*-modulation.

4. Experimental data of Acquisition and Processing Parameters

Table 4.1. Experimental Data of 1, 8-dihydroxy-3-methylanthracene-9, 10-dione for Fig 4.2 and Figs 4.3A-C

Exp. Name	Acq (s)	D1 (s)	SW (ppm)	Data points	NS	Offset (ppm)	FID Res. (Hz)	Zero Filling	LB (Hz)	SSB	WDW
HMQC	0.197	2.0	12.98	2048	2	7.44	5.07	4096	0	2	QSINE
	0.189	-	22.00	840	-	128.5	5.27	2048	0	2	QSINE
r-t pure shift HMQC	0.197	2.0	12.98	2048	2	7.44	5.07	4096	0	2	QSINE
	0.189	-	22.00	840	-	128.5	5.27	2048	0	2	QSINE
r-t pure shift HSQC	0.197	2.0	12.98	2048	2	7.44	5.07	4096	0	2	QSINE
	0.189	-	22.00	840	-	128.5	5.27	2048	0	2	QSINE

For Figures 4.3D-F, the t_1 acquisition time was 20 ms, rest parameters same as above.

Table 4.2. Experimental Data of Glucose

Exp. Name	Acq (s)	D1 (s)	SW (ppm)	Data points	NS	Offset (ppm)	FID Res. (Hz)	Zero Filling	LB (Hz)	SSB	WDW
HMQC	0.169	1.7	2.60	352(F_2)	2	4.1	5.91	512	0	2	QSINE
	0.052	-	38.00	400 (F_1)	-	78.0	19.11	1024	0	2	QSINE
r-t pure shift HMQC	0.169	1.7	2.60	352(F_2)	2	4.1	5.91	512	0	2	QSINE
	0.052	-	38.00	400 (F_1)	-	78.0	19.11	1024	0	2	QSINE
r-t pure shift HSQC	0.169	1.7	2.60	352(F_2)	2	4.1	5.91	512	0	2	QSINE
	0.052	-	38.00	400 (F_1)	-	78.0	19.11	1024	0	2	QSINE

5. Result and discussion

Fig. 4.2A, 4.2B, and 4.2C display the conventional ^1H-^{13}C HMQC, ps-^1H-^{13}C HMQC, and real-time BIRD ps-^1H-^{13}C HSQC spectra of 1, 8-dihydroxy-3-methylanthracene-9,10-dione respectively (Only aromatic signals are shown for clarity) for a longer t_1 acquisition time of 189 ms. The panels e, f, and g, each shows one cross peak expanded from the corresponding peaks in the conventional HMQC spectrum of 4.2A. Similarly, panels e', f', and g' display corresponding expansion from ps-HMQC of 4.2B. Further e'', f'', and g'' are from real-time BIRD ps-HSQC of 4.2C. When compared, singlets along both dimensions of ps-HMQC and ps-HSQC are evident, whereas, in conventional HMQC, unresolved multiplets along both dimensions are observed for each peak. The F_2 projections from Fig. 4.2A, 4.2B, and 4.2C are shown in Fig. 4.3A, 4.3B and 4.3C, which reveals enhanced intensity of the peaks in ps-HMQC (4.3B) and ps-HSQC (4.3C) projections relative to the conventional HMQC (4.3A) as a result of multiplet to singlet conversion. The intensity of different peaks is enhanced by different

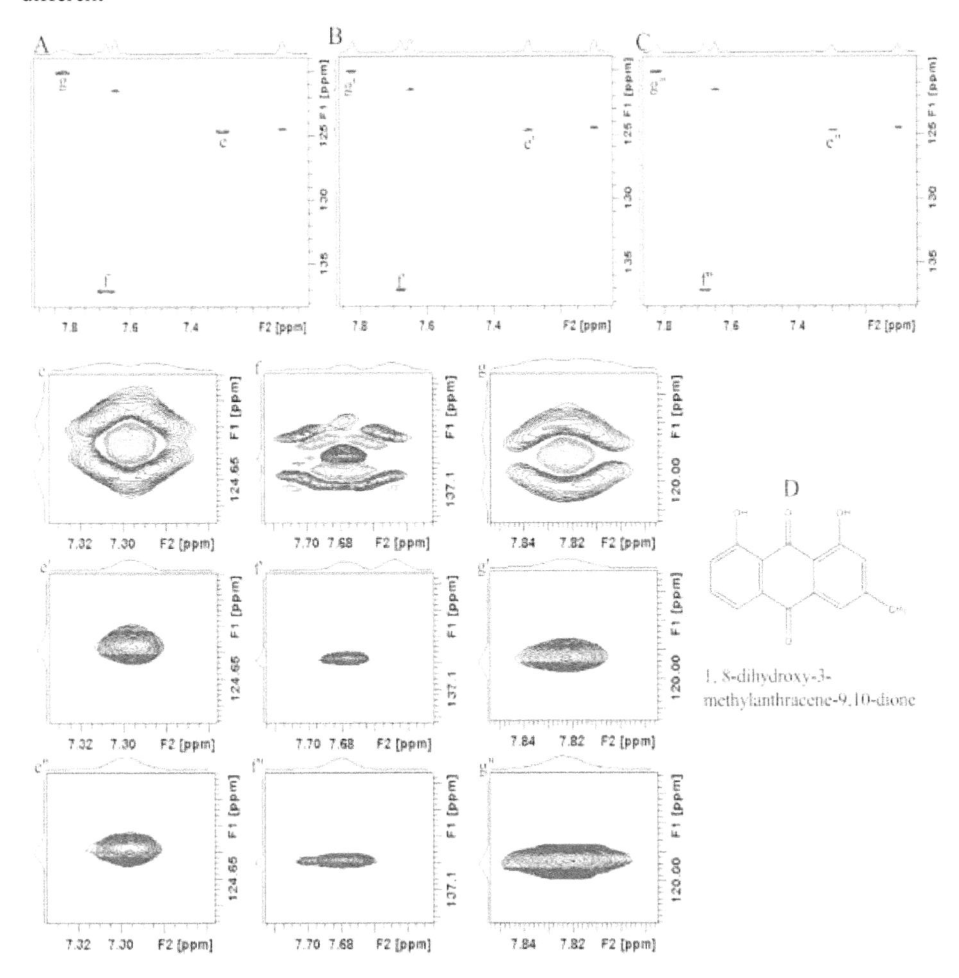

Fig. 4.2. displays spectra from (A) conventional 2D ^{1}H-^{13}C HMQC, (B) ps-^{1}H-^{13}C HMQC, and (C) Real-time BIRD ps-^{1}H-^{13}C HSQC, respectively for the molecule shown in 2D dissolved in CDCl3 (Only aromatic region is shown for clarity, methyl signal is not shown). Panels e, f, and g are extracted portions of the conventional 2D ^{1}H-^{13}C HMQC spectrum from 4.1A. Panels e', f', and g' are extracted portions of the ps-HMQC spectrum from 4.1B. Panels e", f", g" are extracted portions of real-time BIRD ps-HSQC spectrum. All experiments were recorded under the same acquisition and processing parameters. Comparison of the panels e, e', and e" reveals multiplet to singlet conversion in either dimension of ps-HMQC and ps-HSQC. Other panels also show a similar result.

amounts due to the different multiplet patterns involved. For the two doublets, 3.5 and 3.7 times enhanced intensity, and for the triplet, 1.6 times enhancement was observed in ps-HMQC projection (4.3B). The corresponding enhancement factors were 4.1 (d), 4.4 (d), and 2.1 (t) in ps-HSQC (4.3C), which were 10 to 20% higher than ps-HMQC due to the additional t_1 time domain relaxation in the ps-HMQC. Singlets at 7.08 and 7.62 ppm are also enhanced by 1.4 times for ps-HMQC (supported by 1.5 and 1.7 times gain in ps-HSQC) as these are actually broad singlets with four bond unresolved ^1H-^1H couplings between the two aromatic ^1Hs in the methyl containing ring of the molecule shown in Fig. 4.2D.

Thus, for a longer t_1 acquisition time of 189 ms, significant sensitivity enhancement of the peaks was observed for the ps-HMQC and ps-HSQC relative to the conventional HMQC. It is also important to consider the fact that in regular HMQC, evolution of the ^1H-^1H homonuclear couplings during t_1 leads to the creation of antiphase terms which relax faster than inphase terms and this aspect is pre- sent in both conventional HMQC and ps-HMQC, but absent in ps- HSQC, which favors intensity gain in ps-HSQC relative to ps- HMQC. Further, the phase distortions due to antiphase dispersive terms in conventional HMQC (see Fig. 4.2, panels e, f, and g) interferes with the absorptive antiphase terms, which results in a signal loss in the conventional HMQC and this loss is higher for longer t_1 acquisition times. Antiphase dispersive terms are absent in the final ps-HMQC spectrum as coupling gets refocused and does not appear at all in ps-HSQC as there is no ^1H-^1H *J*-modulation along t_1. The loss of signal due to the faster relaxation of antiphase terms and phase distortions are lower for shorter t_1 acquisition time as the buildup of antiphase dispersive terms are less for short t_1 acquisition time. This implies that signal intensity gain will be less when the above spectra are compared for shorter t_1 acquisition time. Therefore, another series of HMQC experiments were recorded for a t_1 acquisition time of 20 ms(much smaller than 189 ms case above), and the F_2 projections of these are shown in Fig. 4.3D, 4.3E, and 4.3F respectively. Inspection of these projections reveals close to two-fold gain in the intensity of the multiplets in ps-HMQC and ps-HSQC. ps-HSQC peaks have a slightly better intensity of the peaks than ps-HMQC. No gain for the singlets (0.9 and 1.1/0.9) as the small long-range couplings do not evolve much for 20 ms. The intensity of the peaks in ps-HMQC and ps- HSQC projections are also comparable as the transverse relaxation due to additional t_1 time domain in ps-HMQC is less for shorter t_1. In practical applications, a t_1 acquisition time of the order of 20 ms is preferred as the experimental time remains reasonable. Further, processing with linear prediction can be used to improve the resolution.

The conventional HMQC, ps-HMQC, and ps-HSQC were also evaluated on glucose for a t_1 acquisition time of 52 ms and are shown in Fig. 4.4 (A, B, C). The peaks marked with a in conventional HMQC spectrum is unresolved due to multiplet structures in both dimensions and being very closely resonating in both frequency domain. These unresolved peaks get resolved in ps-HMQC and ps-HSQC spectral portions marked with a' and a" respectively. Comparison of the F_2 projections in F, G, and H also reveals enhanced sensitivity of the ps-HMQC and ps-HSQC cross-peaks, which are now singlets instead of multiplets. However, one draw- back of ps-HMQC relative to ps-HSQC is apparent when we compare the peaks inside boxes b' and b". The BIRD module fails to decouple geminal ^1H-^1H coupling interaction, and as a result,

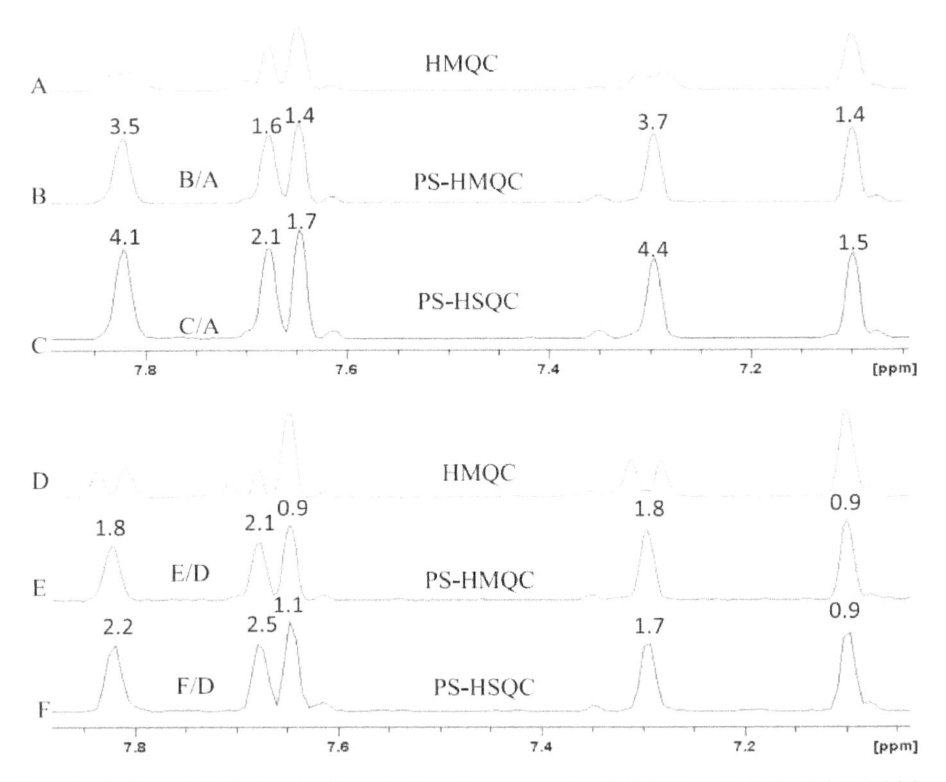

Fig. 4.3. A, B, and C display the F_2 projections from conventional HMQC, ps-HMQC, and ps-HSQC, respectively, from the 4.2D spectra shown in Fig. 4.2A-C. The intensity of the peaks is enhanced many times in ps-HMQC and ps-HSQC projection shown by the numbers by the side of the peaks. All spectra in A, B, and C were recorded for longer t_1 acquisition time of 189 ms, and despite the t_1 time domain being two times longer in

ps-HMQC, significant enhancement occurs due to the reduction of the multiplets to singlets. ps-HSQC displayed a 10–20% higher enhancement than ps-HMQC. D, E, and F display the F_2 projections from conventional HMQC, ps-HMQC, and ps-HSQC, respectively recorded with shorter t_1 acquisition time of 20 ms. When compared to conventional HMQC, enhancement factors in E and F for short t_1 are less than that in B and C for long t_1 as the phase distortions and antiphase contribution to relaxation in regular HMQC is less for short t_1 acquisition time. However, the benefit of real-time decoupling in t_2 remains in ps-HMQC and ps-HSQC irrespective of t_1 acquisition times. Overall, in ps-HMQC and ps-HSQC, enhancements are less for short t_1 acquisition time but much higher for longer t_1 acquisition time when compared to conventional HMQC.

phase distortions due to geminal coupling evolution along t_1 exists, which leads to unresolved peaks in spectral portion b'. There are three peaks in this region (two geminal sites and one methane site), which is clearly resolved in ps-HSQC spectral portion b". The regular HMQC also suffers from the same phase distortions as in ps- HMQC shown in box b.

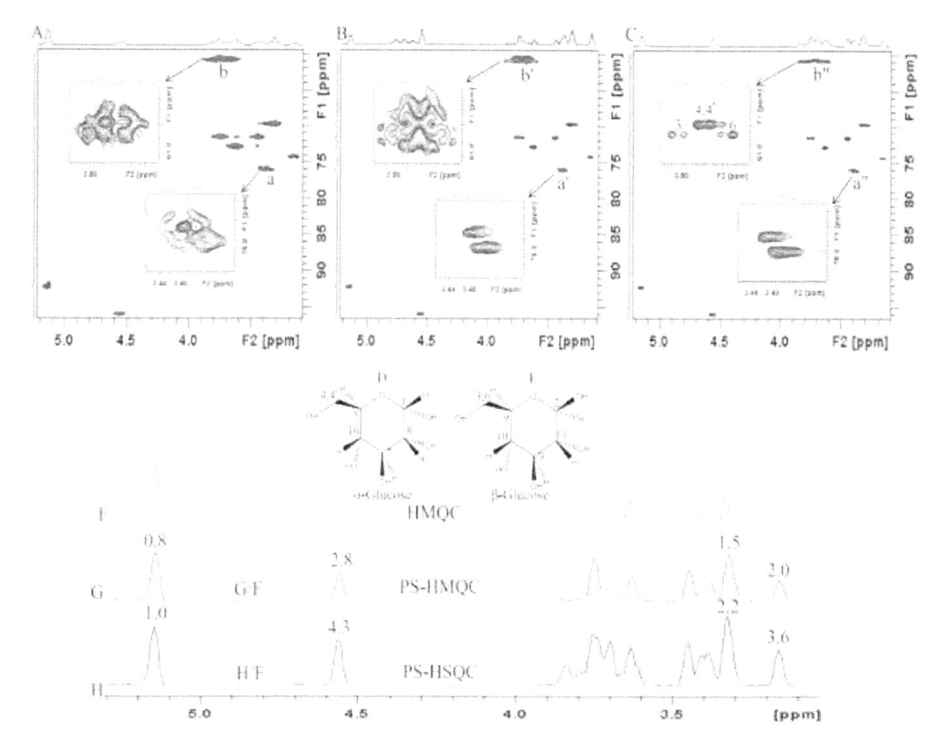

Fig. 4.4. A, B, and C display the conventional HMQC, ps-HMQC, and ps-HSQC spectra, respectively, recorded on a glucose sample. The conventional HMQC spectrum in A displays a peak marked with a, which is unresolved due to multiplet structures in both dimensions. These unresolved peaks get resolved in ps-HMQC, and ps-HSQC spectral portions marked with a' and a" respectively in B and C. Enhanced sensitivity of the ps-

HMQC and ps-HSQC peaks are also apparent from the comparisons of the F_2 projections F, G, and H. Comparison of the boxes b, b' and b" shows that for geminal protons HSQC provides the best resolution. More details are given in the results and discussion.

6. Conclusion

We have developed a way to perform broadband homodecoupling in the indirect dimension of multidimensional experiments, which involves heteronuclear multiple quantum between ^{1}H and ^{13}C. It is demonstrated that BIRD can be used in the direct as well as indirect dimension of the HMQC experiment, which results in the ps-HMQC spectrum. Multiplet to singlet conversion is observed along both dimensions except for diastereotopic site. The method displayed higher sensitivity and resolution compared to conventional HMQC when investigated for short (20 ms) as well as long (190 ms) t_1 acquisition times. When relaxation is not significant, the performance is closer to ps-HSQC. Unlike constant time experiments which suffer from signal loss due to mismatched *J*-couplings, this method of F_1 decoupling using BIRD is robust and independent of the spin system parameters. ^{1}H-^{15}N ps-HMQC experiment can also be designed using BIRD.

While the ps-HSQC experiment remains the best, another useful application of our methodology can be developed for the HMQC-TOCSY experiment. F_1 decoupling using BIRD can be applied along the t_1 dimension of HMQC-TOCSY as in the present work to improve F_1 resolution and overall sensitivity for small molecules. This will be important as the number of cross-peaks is much higher in HMQC-TOCSY than HMQC, and lack of resolution often leads to ambiguity in assignment for crowded spectra. However, real-time decoupling in HMQC-TOCSY using BIRD is not feasible as most of the cross-peaks involves remote correlation to ^{12}C attached protons from ^{13}C attached proton, which are coupled among each other. ZS and PSYCHE can also be utilized instead of BIRD for F_1 decoupling, but this will have further sensitivity penalty multiplied to the natural abundant 1.1% sensitivity of protons. All these aspects are currently under investigation.

7. Declaration of Competing Interest

The authors declare that they have no known competing financial interests or personal relationships that could have appeared to influence the work reported in this paper.

8. Acknowledgments

Upendra Singh thanks the UGC (University Grants Commission) India, for a research fellowship. We thank the SERB (DST) as this work was supported by the extramural research fund provided by the Science and Engineering Research Board under the Department of Science & Technology, Govt. Of India DST No: EMR/2014/001280 (Grant No. SERB/F/6435/2015-16).

9. Future Work

An application of other pure-shift element as perfect-echo (PE) will be implemented in the HMQC in the place of BIRD element in F_1 dimension to refocus homonuclear *J*-scalar coupling leading to overcome in decoupling of geminal protons and enhance the resolution of spectrum but at some cost of sensitivity compare to BIRD pure shift HMQC.

References:

[1] K. Zangger, H. Sterk, Homonuclear broadband-decoupled NMR spectra, J. Magn. Reson. 124 (1997) 486–489.

[2] N.H. Meyer, K. Zangger, Simplifying proton NMR spectra by instant homonuclear broadband decoupling, Angew. Chem. Int. Ed. 52 (2013) 7143– 7146.

[3] J. A. Aguilar, S. Faulkner, M. Nilsson, G.A. Morris, Pure Shift 1H NMR: a resolution of the resolution problem?, Angew. Chem. Int. Ed. 49 (2010) 3901– 3903.

[4] P. Sakhaii, B. Haase, W. Bermel, R. Kerssebaum, G.E. Wagner, K. Zangger, Broadband homodecoupled NMR spectroscopy with enhanced sensitivity, J. Magn. Reson. 233 (2013) 92–95.

[5] S. Glanzer, E. Schrank, K. Zangger, A general method for diagonal peak suppression in homonuclear correlated NMR spectra by spatially and frequency selective pulses, J. Magn. Reson. 232 (2013) 1–6.

[6] G. A. Morris, J.A. Aguilar, R. Evans, S. Haiber, M. Nilsson, True chemical shift correlation maps: A TOCSY experiment with pure shifts in both dimensions, J. Am. Chem. Soc. 132 (2010) 12770–12772.

[7] J. A. Aguilar, A.A. Colbourne, J. Cassani, M. Nilsson, G.A. Morris, Decoupling two-dimensional NMR spectroscopy in both dimensions: pure shift NOESY and COSY, Angew. Chem. 51 (2012) 6460–6463.

[8] K. Zangger, Pure Shift NMR, Prog. Nucl. Magn. Reson. Spectrosc. 86–87 (2015) 1–20.

[9] L. Castanar, T. Parella, Broadband ^{1}H homodecoupled NMR experiments: recent developments, methods and applications, Magn. Reson. Chem. 53 (2015) 399–426.

[10] M. G. Concilio, P. Kiraly, G.A. Morris, Diffusional attenuation during soft pulses: A Zangger-Sterk pure shift iDOSY experiment, J. Magn. Reson. 301 (2019) 85– 93.

[11] J. P. Garbow, D.P. Weitekamp, A. Pines, Bilinear rotation decoupling of homonuclear scalar interactions, Chem. Phys. Lett. 93 (2015) 504–509.

[12] A. Bax, Broadband homonuclear decoupling in heteronuclear shift correlation NMR spectroscopy, J. Magn. Reson. 53 (1983) 517–520.

[13] D. Uhrín, T. Liptaj, K.E. Kövér, Modified BIRD pulses and design of heteronuclear pulse sequences, J. Magn. Reson. 101 (1993) 41–46.

[14] J. A. Aguilar, M. Nilsson, G.A. Morris, Simple proton spectra from complex spin systems: pure shift NMR spectroscopy using BIRD, Angew. Chem. Int. Ed. 50 (2011) 9716–9717.

[15] T. Reinsperger, B. Luy, Homonuclear BIRD-decoupled spectra for measuring one-bond couplings with highest resolution: CLIP/CLAP-RESET and constant- time-CLIP/CLAP-RESET, J. Magn. Reson. 239 (2014) 110–120.

[16] L. Kaltschnee, A. Kolmer, I. Timári, V. Schmidts, R.W. Adams, M. Nilsson, K.E. Kövér, G.A. Morris, C.M. Thiele, ''Perfecting'' pure shift HSQC: full homodecoupling for accurate and precise determination of heteronuclear couplings, Chem. Commun. 50 (2014) 15702–15705.

[17] A. Lupulescu, G.L. Olsen, L. Frydman, Toward single-shot pure-shift solution 1H NMR by trains of BIRD-based homonuclear decoupling, J. Magn. Reson. 218 (2012) 141–146.

[18] L. Paudel, R.W. Adams, P. Király, J.A. Aguilar, M. Foroozandeh, M.J. Cliff, M. Nilsson, P. Sándor, J.P. Waltho, G.A. Morris, Simultaneously enhancing spectralresolution and sensitivity in heteronuclear correlation NMR spectroscopy, Angew. Chem. Int. Ed. 52 (2013) 11616–11619.

[19] A. Verma, B. Baishya, Real-time bilinear rotation decoupling in absorptive mode *J*-spectroscopy: detecting low-intensity metabolite peak close to high- intensity metabolite peak with convenience, J. Magn. Reson. 266 (2016) 51– 58.

[20] Verma, R. Parihar, S. Bhattacharya , B. Baishya, Analyses of complex mixtures by F_1 homo-decoupled diagonal suppressed total correlation spectroscopy, 18 (2017) 3076–3082.

[21] P. Nolis, K.M. Corral, M.P. Trujillo, T. Parella, Broadband homodecoupled time-shared ^{1}H–^{13}C and ^{1}H–^{15}N HSQC experiments, J. Magn. Reson. 298 (2019) 23– 30.

[22] J. D. Haller, A. Bodor, B. Luy a Real-time pure shift measurements for uniformly isotope-labeled molecules using X-selective BIRD homonuclear decoupling, J. Magn. Reson. 302 (2019) 64–71

[23] P. Sakhaii, B. Haase, W. Bermel, Experimental access to HSQC spectra decoupled in all frequency dimensions, J. Magn. Reson. 199 (2009) 192–198.

[24] I. Timari, C. Wang, A.L. Hansen, G.C. dos Santos, S.O. Yoon, L. Bruschweiler-Li, R. Brüschweiler, Real-time pure shift HSQC NMR for untargeted metabolomics, Anal. Chem. 91 (2019) 2304–2311.

[25] L. Castañar, P. Nolis, A. Virgili, T. Parella, Full sensitivity and enhanced resolution in homodecoupled band-selective NMR experiments, Chem. Eur. J. 19 (2013) 17283–17286.

[26] J. Ying, J. Roche, A. Bax, Homonuclear decoupling for enhancing resolution and sensitivity in NOE and RDC measurements of peptides and proteins, J. Magn. Reson. 241 (2014) 97–102.

[27] L. Castañar, J. Saurí, P. Nolis, A. Virgili, T. Parella, Implementing homo- and heterodecoupling in region-selective HSQMBC experiments, J. Magn. Reson. 238 (2014) 63–69.

[28] A. Verma, B. Baishya, Real-time band-selective homonuclear proton decoupling for improving sensitivity and resolution in phase-sensitive *J*-resolved spectroscopy, ChemPhysChem 16 (2015) 2687–2691.

[29] M. Foroozandeh, R.W. Adams, M. Nilsson, G.A. Morris, Ultrahigh-resolution NMR spectroscopy, Angew. Chem. Int. Ed. 53 (2014) 6990–6992.

[30] M. Foroozandeh, R.W. Adams, M. Nilsson, G.A. Morris, Ultrahigh-resolution total correlation NMR spectroscopy, J. Am. Chem. Soc. 136 (2014) 11867– 11869.

[31] V. M. R. Kakita, R.V. Hosur, Non-uniform-sampling ultrahigh resolution TOCSY NMR: analysis of complex mixtures at microgram levels, ChemPhysChem 17 (2016) 2304–2308.

[32] L. Kaltschnee, K. Knoll, V. Schmidts, R.W. Adams, M. Nilsson, G.A. Morris, C.M. Thiele, Extraction of distance restraints from pure shift NOE experiments, J. Magn. Reson. 271 (2016) 99–109.

[33] V. M. R. Kakita, K. Rachineni, R.V. Hosur, Fast and simultaneous determination of ^{1}H–^{1}H and ^{1}H–^{19}F scalar couplings in complex spin systems: Application of PSYCHE homonuclear broadband decoupling, Magn. Reson. Chem. 56 (2018) 1043–1046.

[34] J. A. Aguilar, R. Belda, B.R. Gaunt, A.M. Kenwright, I. Kuprov, Separating the coherence transfer from chemical shift evolution in high-resolution pure shift COSY NMR, Magn Reson Chem. 56 (2018) 969–975.

[35] N. Giraud, L. Béguin, J. Courtieu, D. Merlet, Nuclear magnetic resonance using a spatial frequency encoding: application to J-edited spectroscopy along the sample, Angew. Chem. Int. Ed. 49 (2010) 3481–3484.

[36] M. Foroozandeh, R.W. Adams, P. Kiraly, M. Nilsson, G.A. Morris, Measuring couplings in crowded NMR spectra: pure shift NMR with multiplet analysis, Chem. Commun. 51 (2015) 15410–15413.

[37] D. Sinnaeve, M. Foroozandeh, M. Nilsson, G.A. Morris, A general method for extracting individual coupling constants from crowded 1H NMR spectra, Angew. Chem. Int. Ed. 128 (2016) 1102–1105.

[38] P. Kiraly, M. Foroozandeh, M. Nilsson, G.A. Morris, Anatomising proton NMR spectra with pure shift 2D J-spectroscopy: a cautionary tale, Chem. Phys. Lett. 683 (2017) 398–403.

[39] J. E. H. Pucheta, D. Pitoux, Claire M. Grison, S. Robin, D. Merlet, David J. Aitken, N. Giraud, J. Farjon, Pushing the limits of signal resolution to make coupling measurement easier, Chem. Commun. 51 (2015) 7939–7942.

[40] Y. Lin, Q. Zeng, L. Lin, Z. Chen, P.B. Barker, High-resolution methods for the measurement of scalar coupling constants, Prog. Nucl. Magn. Reson. Spectrosc. 109 (2018) 135–159.

[41] S. Berger, A quarter of a century of SERF: The progress of an NMR pulse sequence and its application, Prog. Nucl. Magn. Reson. Spectrosc. 108 (2018) 74–114.

[42] D. Sinnaeve, M. Foroozandeh, M. Nilsson, G.A. Morris, A general method for extracting individual coupling constants from crowded ^{1}H NMR spectra, Angew. Chem. Int. Ed. 55 (2016) 1090–1093.

[43] A Verma, B Baishya, Real-Time Band-Selective Homonuclear Proton Decoupling for Improving Sensitivity and Resolution in Phase-Sensitive *J*-Resolved Spectroscopy, 16 (2015) 2687–2691.

[44] I. Timári, L. Kaltschnee, A. Kolmer, R.W. Adams, M. Nilsson, C.M. Thiele, G.A. Morris, K.E. Kövér, Accurate determination of one-bond heteronuclear coupling constants with "pure shift" broadband proton-decoupled CLIP/ CLAP-HSQC experiments, J. Magn. Reson. 239 (2014) 130–138.

[45] J. Dittmer, G. Bodenhausen, Quenching Echo Modulations in NMR Spectroscopy, Chem Phys Chem. 7 (2006) 831–836.

[46] B. Baishya, T.F. Segawa, G. Bodenhausen, Apparent transverse relaxation rates in systems with scalar-coupled protons, J. Am. Chem. Soc. 131 (2009) 17538– 17539.

[47] T. F. Segawa, B. Baishya, G. Bodenhausen, Transverse relaxation of scalar- coupled protons, Chem Phys Chem. 11 (2010) 3343–3354.

[48] D. Carnevale, T.F. Segawa, G. Bodenhausen, Polychromatic decoupling of a manifold of homonuclear scalar interactions in solution-state NMR, Chem.-A Eur. J. 18 (37) (2010(2012)) 11573–11576.

[49] K. Takegoshi, K. Ogura, K. Hikichi, A perfect spin echo in a weakly homonuclear *J*-coupled two spin 1/2 system, J. Magn. Reson. 84 (1989) 611–615.

[50] P.C. van Zijl, C.T.W. Moonen, M.J. Von Kienlin, Homonuclear J-refocusing in echo spectroscopy, J. Magn. Reson. 89 (1990) 28–40.

[51] R.W. Adams, C.M. Holroyd, J.A. Aguilar, "Perfecting" WATERGATE: clean proton NMR spectra from aqueous solution, M. Nilsson and G. A. Morris, Chem. Commun. 49 (2013) 358–360.

[52] J. A. Aguilar, M. Nilsson, G. Bodenhausen, Spin echo NMR spectra without *J*-modulation, G. A. Morris, Chem. Commun. 48 (2012) 811–813.

[53] B. Baishya, C.L. Khetrapal, K.K. Dey, "Perfect echo" HMQC: sensitivity and resolution enhancement by broadband homonuclear decoupling, J. Magn. Reson. 234 (2013) 67–74.

[54] B. Baishya, C. L. Ketrapal, "Perfect echo" INEPT: more efficient heteronuclear polarization transfer by refocusing homonuclear *J*-coupling interaction, J. Magn. Reson. 242 (2014) 143–154.

[55] L. Castañar, E. Sistare, A. Virgili, R.T. Williamson, T. Parella, Suppression of phase and amplitude J(HH) modulations in HSQC experiments, Magn. Reson. Chem. 53 (2015) 115–119.

[56] B. Baishya, A. Verma, Elimination of Zero-Quantum artifacts and sensitivity enhancement in perfect echo based 2D NOESY, J. Magn. Reson. 252 (2015) 41–48.

[57] J. Ilgen, L. Kaltschnee, C.M. Thiele, A pure shift experiment with increased sensitivity and superior performance for strongly coupled systems, J. Magn. Reson. 286 (2018) 18–29.

[58] A. Bax, A.F. Mehlkopf, J. Smidt, Homonuclear broadband decoupled absorption spectra, J. Magn. Reson. 35 (1979) 167–169.

[59] C. Griesinger, O.W. Soerensen, R.R. Ernst, Two-dimensional correlation of connected NMR transitions, J. Am. Chem. Soc. 107 (1985) 6394–26396.

[60] W. P. Aue, J. Karhan, R.R. Ernst, Homonuclear broad band decoupling and two-dimensional J-resolved NMR spectroscopy, J. Chem. Phys. 64 (1976) 4226– 4227.

[61] L. Mueller, Sensitivity enhanced detection of weak nuclei using heteronuclear multiple quantum coherence, J. Am. Chem. Soc. 101 (1979) 4481–4484.

[62] Ad Bax, R.H. Griffey, B.L. Hawkins, Correlation of proton and nitrogen-15 chemical shifts by multiple quantum NMR, J. Magn. Reson. 55 (1969) 301–315.

[63] P. Schanda, B. Brutscher, Very Fast two-dimensional NMR spectroscopy for real-time investigation of dynamic events in proteins on the time scale of seconds, J. Am. Chem. Soc. 127 (2005) 8014–8015.

[64] A. Ross, M. Salzmann, H. Senn, Fast-HMQC using Ernst angle pulses: An efficient tool for screening of ligand binding to target proteins, J. Biomol. NMR 10 (1997) 389–396.

[65] S. Grzesiek, A. Bax, Spin-locked multiple quantum coherence for signal enhancement in heteronuclear multidimensional NMR experiments, J. Biomol. NMR 6 (1995) 335–339.

[66] H. Kuboniwa, S. Grzesiek, F. Delaglio, A. Bax, Measurement of HN-Ha *J*- couplings in calcium-free calmodulin using new 2D and 3D water-flip-back methods, J. Biomol. NMR 4 (1994) 871–878.

[67] H. Ponstingl, G. Otting, Rapid measurement of scalar three-bond 1HN-1Ha spin coupling constants in 15N-labelled proteins, *J.* Biomol. NMR 12 (1998) 319– 324.

[68] A. Bax, L.E. Kay, S.W. Sparks, D. A. Torchia, Line narrowing of amide proton resonances in 2D NMR spectra of proteins, J. Am. Chem. Soc. 1 (1989) 408–409.

[69] W. Kozminski, A pure-phase homonuclear *J*-modulated HMQC experiment with tilted cross-peak patterns for an accurate determination of homonuclear coupling constants, J. Magn. Reson. 141 (1999) 185–190.

[70] A. L. Davis, J. Keeler, E.D. Laue, D. Moskau, Experiments for recording pure-absorption heteronuclear correlation spectra using pulsed field gradients, J. Magn. Reson. 98 207–2 (1992) 16.

[71] J. R. Tolman, J. Chung, J. H. Prestegard, Pure-phase heteronuclear multiple- quantum spectroscopy using field gradient selection, J. Magn. Reson. 98 (1992) 462–1461.

[72] L. E. Kay, P. Keifer, T. Saarinen, Pure absorption gradient enhanced heteronuclear single quantum correlation spectroscopy with improved sensitivity, J. Am. Chem. Soc. 114 (1992) 10663–10665.

[73] G. Zhu, X. Kong, K. Sze, Gradient- and sensitivity-enhanced heteronuclear multiple-quantum correlation spectroscopy, J. Magn. Reson. 135 (1998) 232– 235.

[74] T. Parella, Pulse program catalogue: I. 1D & 2D NMR experiments, Bruker Biospin

[75] P. Kiraly, M. Nilsson, G.A. Morris, Practical aspects of real-time pure shift HSQC experiments, Magn Reson Chem. 56 (2018) 993–1005.

[76] M.A. Smith, H. Hu, A.J. Shaka, Improved broadband inversion performance for NMR in liquids, J. Magn. Reson. 151 (2001) 269–283.

[77] M. Garwood, L.D. Barre, The return of the frequency sweep: designing adiabatic pulses for contemporary NMR, J. Magn. Reson. 153 (2001) 155–177.

[78] E. Kupcˇe, R. Freeman, Compensation for spin-spin coupling effects during adiabatic pulses, J. Magn. Reson. 127 (1997) 36–48.

[79] M.H. Levitt, R. Freeman, NMR population inversion using a composite pulse, J. Magn. Reson. 33 (1979) 473–476.

Lightning Source UK Ltd.
Milton Keynes UK
UKHW020702060223
416538UK00014B/1012